Process Control

A Primer
For The Nonspecialist
And The Newcomer

George Platt

INSTRUMENT SOCIETY OF AMERICA

PROCESS CONTROL — A PRIMER FOR THE NONSPECIALIST AND THE NEWCOMER

Instrument Society of America
67 Alexander Drive, P.O. Box 12277
Research Triangle Park, NC 27709

Library of Congress Cataloging in Publication Data

Platt, George, 1920–
    Process control: a primer for the nonspecialist and the newcomer

    George Platt.
    p. cm.
    Bibliography: p.
    Includes index.
    1. Process control.   I. Title.
TS156.8.P53 1988
670.42-dc19   88-15459 CIP
ISBN 1-55617-096-3

The Instrument Society of America wishes to acknowledge the cooperation
of those manufacturers, suppliers, and publishers
who granted permission to reproduce material herein.
The Society regrets any omission of credit that may have occurred
and will make such corrections in future editions.

Second Printing, 1991

Dedication

To Sarah —
my wife, friend, and helpmeet

Contents

Preface

Who Should Read This Book?

Are you an executive with overall responsibility for the production, profitability, and safety of a process plant or other industrial plant? Have you moved up the company ladder, leaving behind much of your technical background? Or are you, perhaps, a whiz in sales or finance but never got into the technical side of things?

What do you do when your plant supervisor asks you to approve a project that requires an important addition or change to the plant instrument system? The project may involve obtaining and spending large sums of money, bearing the costs of downtime, complying with governmental pollution regulations, safety requirements, and so on. Does the plan seem technically practical? You could rely wholly on your technical people, but wouldn't a little knowledge of how instruments control a plant and of the technical lingo help you to make a more intelligent evaluation of the project?

This book surveys the field of process instrumentation and opens the door to understanding its principles and uses. It describes the elements of process control simply and with homey examples so that a non-technical person can understand its essential principles. It also describes the basics of some common types of instruments. The book discusses the importance of process control and its place in the overall plan for managing a company. It touches on engineering and managerial considerations for instrument systems.

Reading the book will not make you a competent instrument specialist, but you will no longer be a total stranger in a strange land. At least you will be able to ask more of the right questions. You may also find the subject very interesting.

This book may also be used in technical schools as an introduction to the ABCs of process control. The book may be useful to people — such as salesmen for machinery systems or electrical or safety engineers — whose concern with process instrumentation is incidental or passing.

The information in this book applies to a wide spectrum of industries, including:

Architect-engineering

Industrial control

Process: food; metal refining and manufacturing; power generation; chemical; rubber; pharmaceutical; petroleum; glass; textile; paper; waterworks; sewage treatment; heating, ventilating, and air conditioning

1

Introduction: What is Meant by a Process?

1-1. WHAT IS A PROCESS?

A *process* is any operation or sequence of operations involving a change in the substance being treated. [1][1] Examples of processes include the following:

- A change of energy state, such as from hot to cold, or liquid to gas.
- A change of composition, as occurs in a chemical reaction or in mixing different materials.
- A change of dimension, as in grinding coal.

This book considers *process* in a broad sense because the applications of instrument principles are widespread. A process may be something as elaborate as deriving gasoline from the very complex mixture of chemicals in crude oil, or something as simple as pumping water from one place to another. It may be heating a house to a

desired temperature; pasteurizing milk; keeping an airplane at a constant speed, direction and elevation; controlling a group of elevators in a building; operating an artificial heart; canning food; tracking a star by telescope; tracking a ship by sonar or radar. For all these processes, certain universal principles of measurement and control apply though the hardware and techniques may differ greatly. Also, the titles given to the people who supervise or direct the process may vary widely: plant operator, hospital nurse, airplane pilot, astronomer, radar technician, and many more.

This book focuses on instrumentation as used in industry, particularly the so-called process industries, which include those for making steel, treating foodstuffs, making chemicals, refining mineral ores, refining petroleum, etc. The hardware mentioned in this book is typical of what these industries use.

Each process has a number of properties that may vary. Examples of properties are pressure, temperature, level, flow rate, acidity, color, quantity, viscosity, and many others. Each of these properties is known as a *process variable*. [2] The changing values of the variables

---

[1] Numbers in brackets are reference numbers. A list of references follows the appendixes.

may be measured and sent to remote locations by means of signals. The measurements may be read, used for control, or stored.

1-2. WHAT ARE PROCESS INSTRUMENTS?

A *process instrument* is a device used directly or indirectly to perform one or more of the following three functions:

• *Measurement:* To measure is to determine the existence or magnitude of a variable (see Chapter 3). Measuring instruments include all devices used directly or indirectly for this purpose. [3] Measurement systems may include auxiliary instruments to provide readings or alarms, to calculate derived values, or to perform other functions.

• *Control:* To control is to cause a process variable, known as the *controlled variable*, to be maintained at a specified value, or within specified limits, or to be altered in a specified manner. For example, room temperature may be controlled by a thermostat to stay constant. A controller requires and commands another device, which is usually a *final control element* (see "Manipulation"). An automatic controller may be looked at as an automatic brain that has no muscle. Different types of controllers may operate automatically, or by manual adjustment, or either way (see Chapter 4). [4]

• *Manipulation:* To manipulate is to cause a final control element to directly change a process variable in order to achieve control of another process variable. [5] For example, the above thermostat operates a fuel valve, which manipulates the fuel flow. The final control element is the soldier, the muscle, that carries out the orders of the commander, the controller (see Section 2-1).

The above three function categories are commonly referred to in the abbreviated form, *measurement and control*. In this descriptive term, the manipulation function is absorbed into the control function because both functions are involved in the corrective action to control the process. Nevertheless, it is important to be aware of the differences in function between control and manipulation.

The term *process instrumentation* covers the above categories of instruments, but, generally, it also includes related accessory instruments, instrument piping, and instrument wiring, if they are required. It excludes power supplies and other gadgetry, such as anti-freeze devices, that merely enable the instrumentation to function.

1-3. WHY ARE PROCESS INSTRUMENTS NEEDED?

The history of mankind is tied intimately to the use of instruments. The measurement of land boundaries and distances; of the amount of periodic rise and fall of the Nile River and ocean tides; of time (by observing heavenly bodies); of weight of cattle, precious metals, etc. — all these measurements in ancient times involved the use of primitive instruments. The age of automatic control began in 19th-century England with James Watt's invention of a governor to control the speed of a steam engine. This evolved into the modern universe of instruments that have an almost limitless variety of function, construction, and purpose.

Now we are in the computer age with its general-purpose computers and special-purpose computers that are being more and more hybridized with and paired with traditional instruments for measurement and control. The capabilities of instrument systems have been steadily increasing with resulting economic benefits to industry. These instrument capabilities include the following:

• *Range and Accuracy of Measurement.* Instruments can measure dimensions over a range from the submicron to light-years; they can measure time intervals to less than a billionth of a second, weights from less than a millionth of a pound up to many tons — far beyond what any human can achieve. Extremely high accuracy is now commonly provided by automatic chemical analyzers that can measure chemical composition to parts per billion. Only through instruments can we have reliably high-purity chemicals and metal alloys and know how good or bad our air and water are.

• *Process Operability.* Some processes would not be practical or even possible if it were not for modern instrumentation. The whole field of nuclear power is a prime example of an industry that could not exist if it were not for the instruments that measure radioactivity. No one among us has any physical sensation when he has a dental x-ray or sits in his living room and receives natural radiation from the walls, but instruments exist that measure such things, and these instruments are vital to the operation of a nuclear power plant.

How would we measure the flow of catalyst for making gasoline in a fluidized-bed catalytic cracker if not for instruments? Without automatic instruments for measurement and control, could we get the exact mix of colors in commercial cans of paint filled by the thousands in a paint factory? Skyscrapers would not be practical without elevators and their control systems to transport people and goods to the heights.

• *Process Quality.* We take high quality for granted for so many products in our technological world. How many hundreds of millions of gallons of milk do we pasteurize properly every year for the health of our people? The pasteurizing process requires the milk to be held at an elevated temperature for some minimum time to kill harmful bacteria. At the same time, the temperature must not be too high and the heating time not too long or else too much nutritive value will be lost. Then the milk is checked for purity and is stored at a controlled low temperature to maintain its quality. To maintain this quality in large-scale production would not be possible without modern measuring and control instruments.

Suppose my car engine requires 88-octane gasoline and yours needs 92 octane. The oil refineries could choose to make only the higher-grade gasoline, which would be fine for your car and mine, but I would then have to pay the price for a better gasoline than my car needs. To satisfy both customers, the refineries produce both grades accurately through the use of sophisticated petroleum refining equipment and instrument systems. Thus my car can have an economical fuel on which it runs well, and your "jazzy" car can have the more powerful fuel it needs.

• *Process Reliability and Safety.* The results might be most unfortunate if the processes discussed above did not operate properly and safely, and we might see frequent outbreaks of intestinal infections from improperly pasteurized milk and widespread dissatisfaction because of automobile engine knock and poor gasoline mileage.

Nuclear power plants are very complex. To enable all their nuclear systems to work well and safely, a staggering number of overlapping instrument systems and subsystems are installed to provide the high reliability that is required in the United States and other countries. All nuclear safety-related instrumentation is manufactured and installed to exceptional standards of conservatism and quality that far exceed what is normal for run-of-the-mill commercial instruments, which are themselves generally of very high quality. Further discussion of reliability is in Section 10-2.

An example of improving plant availability and reliability by using instruments to reduce the servicing frequency of process equipment follows: Feedwater supplied to steam generators, especially large ones, should be of extremely high purity, close to that of distilled water. The purpose is to minimize the corrosion of steam generator tubes and the buildup of mineral scale inside the metal tubes. Corrosion eats away and thins the tube walls, leading to tube leakage. The scale hinders the transfer of heat from hot combustion gases to the water, thereby increasing the fuel usage by the steam generator; this incidentally lowers the operating efficiency and increases the operating cost. The scale also causes the metal tubes to operate hotter, thus weakening the metal and aggravating the corrosion. Continuous chemical analyzers are routinely used to keep the feedwater adequately pure to protect the generator.

Another aspect of reliability is the fact that instruments do not have "people" problems; the instruments are more trustworthy. (For additional discussion of reliability, see Section 10-2.)

• *Process Efficiency.* The efficiency of a process may be very sensitive to how closely the process variables are controlled. Efficiency relates to how much of a desired output can be obtained from a given input to the process. For example, coal containing a certain amount of thermal energy is put into a steam generator and burned. A corresponding amount of steam theoretically would be generated if the plant operated at 100 percent efficiency. But no plant comes close to this ideal because of physical limitations and operating difficulties, including weaknesses of the plant operators. If some of the human weaknesses can be eliminated or reduced by using instruments, the plant efficiency goes up. Higher efficiency means higher productivity and lower cost.

A high efficiency requires not only a good design of the process and the process equipment but also the right instruments to make the process work the way it is supposed to.

• *Labor Cost.* Many plant measurement or control functions that are performed by people can be performed at least as well by automatic instruments. The payback for the cost of instruments, including the cost of servicing the instruments from time to time, often justifies replacing people by instruments. The historical trend has been for process plants to have more and more instruments, both for their comparative economic advantage over people and for their superhuman capabilities.

2

How Instruments Work Together

2-1. THE SEQUENCE OF FUNCTIONS IN A LOOP

In Section 1-2, we stated that there are three general classes of instrument functions that are used in single instruments or in combinations of instruments in order to measure, control, and manipulate the process. All instruments perform one or more of these functions. A combination of instruments or functions that are interconnected to measure or control a process variable is known as a *loop*. [6] Figure 2-1 shows a block diagram for a simple instrument loop to control the speed of an automobile; the control system is known as *cruise control*.

The process is the car engine in operation. Our aim is for the car to automatically keep to a speed that we choose. There are technical variations, but we will assume that the engine speed is an index of car speed, meaning that we will consider the car speed to be constant if the engine speed is constant no matter whether we are traveling uphill, downhill, or on the level.

To control the speed, the driver of the car manually sets a controller to maintain a speed of, for example, 45

miles per hour. The actual speed is continuously measured by a speed sensor, which sends a signal to inform the controller what the speed is. The controller then decides whether the current speed is too high or too low. If too low, the controller sends a signal to a fuel valve to open further to permit the fuel flow to the engine to increase; if too high, the controller commands the valve to be open less; if just right, nothing changes. In any case, the engine speeds up or slows down, if and as required. By continuously measuring, controlling, and manipulating, the cruise control system keeps the car moving at a constant speed.

The figure shows how the sensor does nothing but measure and send out information. The sensor is also known as a *primary element*. The controller does nothing but study the process situation, comparing what is with what should be, and then sends out a command. The final control element does nothing but manipulate the fuel flow. (Here is where it is usually necessary to supply physical force to overcome process forces in order to give good control. In process work, the final control element sometimes applies great force in doing its job.) The engine speed is the controlled variable; fuel flow is the manipulated variable.

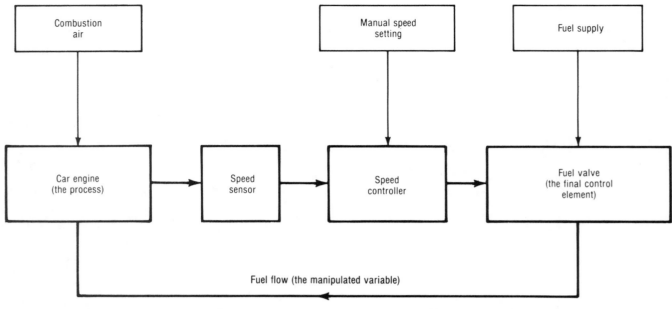

Arrowheads show the direction of flow of material or information.   Cruise control automatically keeps the car speed constant.
The elements of the process control loop are shown with heavy lines.   The fuel flow is manipulated to control the speed.

Figure 2-1. Automobile Cruise Control

An example of another control loop, taken from industry, is shown in Figure 2-2. A constant volume of water is stored in a tank to provide a reserve for process equipment that withdraws water at fluctuating rates. Automatic level control is used to keep the tank level high and to avoid overflowing the tank. A sensor measures the water level and sends to a transmitter a signal that corresponds to the level. A *transmitter* is an instrument that passes on the information that it receives from a sensor; the sensor may or may not be part of the transmitter. [57] The transmitter then sends a measurement signal to a controller that directs the control valve, which, in turn, manipulates the water supply flow to the tank as required to keep the level constant. The controlled variable is water level; the manipulated variable is the flow of water to the storage tank.

The information sent out by the transmitter can be used by instruments other than the controller. Figure 2-2 shows how this same information can be used for other instruments, such as indicators, alarms, and so on. These instruments are not part of the control loop, though they are part of the measurement loop.

2-2. COMBINING FUNCTIONS IN A SINGLE INSTRUMENT

In order to measure or control a process variable, we may need only one instrument in some cases, but, very often, we need a group of instruments that work together, as in Section 2-1.

A household thermometer both senses the temperature and indicates the temperature; it is a complete measuring system, a complete loop. A common industrial loop uses two instruments to provide a temperature measurement to be read at a distant location. These loops are shown in Figure 2-3.

In another case, the level sensor and the level transmitter of Figure 2-2 are frequently separate instruments, as shown, but they can be joined in a single instrument. Thus, we often find many different combinations of sensing, transmitting, controlling, automatic switching, indicating, and other functions in single instruments, such as temperature recording controllers, flow indicating switches, and many others.

The individual functions are not always obvious when their hardware is in one physical assembly, yet the functions are there. By understanding the basic functions, we can better understand how such assemblies work. As an example, consider a household electric iron, which is a single device that is a complete control loop in itself. The iron is a regulator that maintains a constant iron temperature that is adjustable for ironing cotton, nylon, or other fabrics. Its use requires human guidance and human energy, but that is apart from its built-in function of controlling temperature automatically. It does this by measuring its own temperature, providing a corrective command if the temperature is not what the operator has set it to be, and closing or opening an electric switch. On low temperature the switch closes, electricity flows through an electrical heating element, the temperature rises, gets too high; the switch opens, the flow of electricity stops, the temperature gets too low; and the on-off cycle starts again.

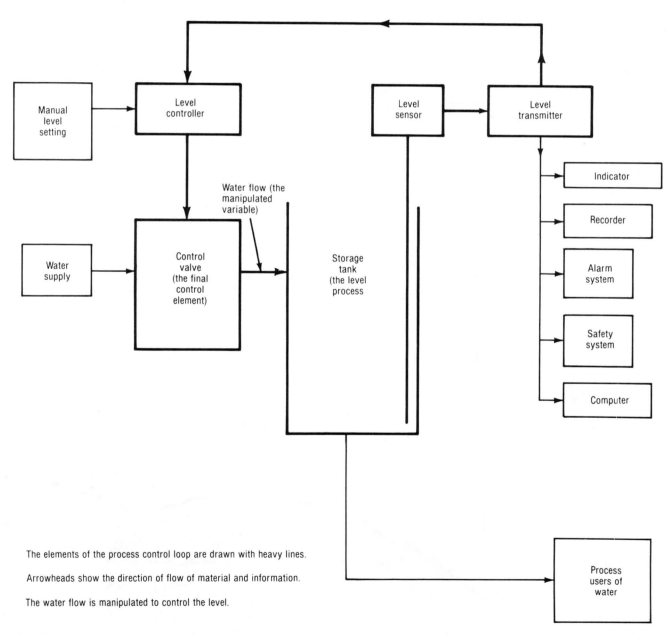

The elements of the process control loop are drawn with heavy lines.

Arrowheads show the direction of flow of material and information.

The water flow is manipulated to control the level.

Figure 2-2. Control of Water Level in a Tank

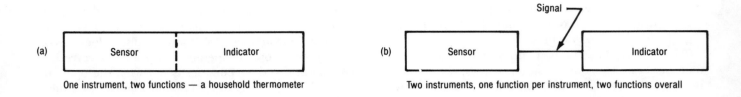

Figure 2-3. Two Loops to Indicate Temperature

3

How Measurements Are Made

3-1. GENERAL PRINCIPLES OF MEASUREMENT

3-1-1. Direct versus Inferred Measurements

There are two general modes of measurement: *direct* and *inferred.*

● *Direct Measurement.* If we want to know whether it is windy outside, we can go outside and feel whether there is a wind. This is a *direct* measurement. In the case of instruments, there are flow meters, such as a household water meter or a gasoline dispenser at an automobile service station, that provide a direct reading of the fluid volume that has passed through.

● *Inferred Measurement.* We can determine whether there is a wind outside by going to a window and seeing whether our flag is waving, or whether the tree branches are swaying, or how rough the water is in the lake. These are *inferred* measurements; they depend on our sensing the effects of the wind instead of the wind itself. The Beaufort scale, which sailors use to judge the strength of

winds, is based on such effects; it covers the range from calm to hurricane. This scale is named for the 18th–19th century British admiral, Sir Francis Beaufort. Similarly, some flow meters determine fluid volume inferentially by sensing over a period of time the pressure effects that are caused by the flow. (For flow sensors, see Section 3-5.)

In general, an inferred measurement senses not the physical property that we want to know about but, instead, an effect created by that property.

3-1-2. Range and Span

Each measured variable and each measuring instrument has a selected range. *Range* is defined by a lower limit and an upper limit of variation or usefulness. [7] The upper limit is known as *full scale.* For example, suppose we manufacture a household thermometer for outdoor use where the temperature may vary between –20°F and +115°F. To provide some margin at the top and bottom, we extend the indicator scale on the thermometer so that it can measure –40 to +130°F. The *design temperature range* is then –20 to +115°F, and the *thermometer range* is –40 to +130°F.

Span is the algebraic difference, the numerical distance, between the lower and upper limits of a range. [8] For our thermometer, the span is +130°F – (–40°F) = 170°F (see Figure 3-1). An automobile speedometer having a range of 0 to 110 miles per hour has a span of 110 miles per hour.

A range is always defined by two numbers, a span by one number.

3-1-3. *Static versus Dynamic Measurements*

We have a pot of warm water on a table and we want to measure the water temperature. We put a thermometer in the water, and we wait a few minutes for the thermometer to be warmed by the water. Then we take a reading, which is easy to do because the water temperature has not changed.

Now we put the pot on a stove and turn the heat on to maximum. We find it difficult to read the temperature because the water temperature is continuously rising. Right now the temperature is 138°F, a moment later it is 139°F, then 140°F, and so on until the water boils. While the water temperature is changing, the thermometer responds but cannot keep up because it is chasing a moving target. When the water temperature stops changing, then the thermometer will be able to catch up.

When the pot of water was not being heated and the water temperature was constant, the measurement was made under *static* conditions, also known as *steady-state* or *equilibrium* conditions. After the thermometer reached the initial water temperature, a good reading could be taken. But when the water was being heated and its temperature rose, the conditions were *dynamic*, also known as *unsteady state*. At a given moment when the water and thermometer temperatures were changing, their difference was the *dynamic error*. These two cases are shown in Figure 3-2.

The term *dynamic* is relative. Some sensors are very quick to respond to changes and are said to have a fast response and a short response time; some have a slow response and a long response time. We might have little difficulty in tracking — keeping up with the water-temperature changes — if we used a fast-response sensor or if we heated the water slowly; the dynamic error would be small.

Figure 3-3 shows how an abrupt change of instrument input from its initial steady state to its final steady state causes a gradual change of instrument output. The abrupt change, which theoretically is done instantaneously, is known as a *step change*. A standard way of describing the response time for a simple system is by a *time constant*, which is based on a 63.2 percent change of output caused by a step change of input. [9] For example, the time constant of a given instrument is stated to be six seconds, which means that, for a step change of input, the output will reach 63.2 percent of its ultimate change in six seconds.

In comparing the response times for two different instruments, we have to be sure that the times are stated in the same way. When this is not done, it causes difficulties in evaluating the relative performance of different instruments for purchase. Table 3-1 shows situations that may arise.

Figure 3-1. Instrument Range and Span

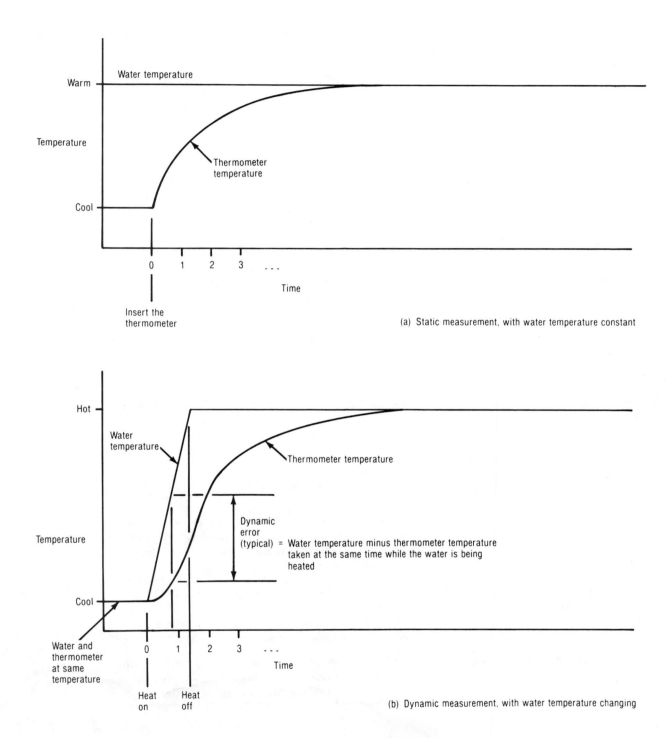

Figure 3-2. Temperature Measurement under Static and Dynamic Conditions

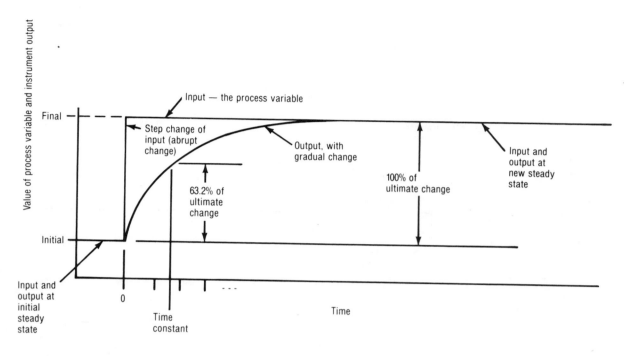

Figure 3-3. Instrument Response to Abrupt Change of Input in a Simple System

Table 3-1. Statements of Instrument Response Time

Instrument Response Time	Comment
Time constant is 6 seconds.	This statement is suitable for making direct comparison with other time constants, all of which relate to a 63.2 percent change.
Response time is 9 seconds for 90 percent change.	We cannot say whether this response is faster or slower than that of a 6-second time constant. The greater time of 9 seconds relates to a change greater than 63.2 percent. Mathematical analysis is required.
Response time is 4 seconds.	Comparison with response statements for other instruments is not possible because the 4 seconds are not related to any specific amount of change.

3-1-4. Instrument Calibration

We speak of an instrument being calibrated. To *calibrate* a device means to determine its outputs that correspond to a series of inputs to the device. The data thus obtained are used to:

- Determine the locations at which scale graduations are to be placed.
- Adjust the instrument output to the desired value or values. For example, a temperature transmitter for measuring a range of 0 to 200°F requires an output signal range of 4 to 20 milliamperes. The instrument is calibrated accordingly.
- Ascertain the error by comparing the actual output against what the output should be. [10]

3-1-5. Truth in Measurement

All measurements are imperfect because the instruments that make the measurements are imperfect. There are limits to how finely any measurement can be made. The closest approach to the TRUE value of a measured property — for example, the height of a table — is obtained only by using the best measuring tool available, the best measuring technique, and taking the average of many measurements. Even the ultra-accurate measurements of the U. S. Bureau of Standards are imperfect. Following are terms that describe how true are measurements and how much faith we can place in them.

- *Accuracy.* Assume the room we are in happens to have an actual temperature of 72.128641...°F even though we do not know this. We look at the room thermometer, which shows a temperature of a hair over 72°F, and we announce that the temperature is 72 degrees. We cannot read the exact temperature no matter how careful we are. We could use a laboratory thermometer and read the temperature as 72.13°F or even 72.129°F, but there would still be a residual error. We never reach the absolute truth; we only approach it more or less depending on how good our measuring instrument is.

A measurement error equals the reading, the apparent value, minus the true or ideal value. [11] In practical terms, let us say that our actual room temperature is what a very accurate thermometer says it is: 72.129°F. The error is then 72°F minus 72.129°F, or –0.129°F. A negative error means that the reading is less than the true value; a positive error means that the reading is higher than the true value. The error in a given case may or may not be important. For us sitting in the room, who cares if we think the temperature is 72°F instead of 72.129°F?

A peculiarity of engineering terminology is that an error, which is the *inaccuracy*, is usually called *accuracy*. This "accuracy" is –0.129/72 = –0.0018 of reading = –0.18% of reading.

The way that comparisons are stated can cause great misunderstanding when applied to accuracy or other things. For example, you put some money into a bank account, and you receive six percent interest. Later you switch to a different kind of account and you receive eight percent interest. You tell me that your interest rate has increased by two percent. I say it increased by 33-1/3 percent. Can we both be right?

The problem here is that each of us is using a different and unstated base of reference. If you will say that the rate has increased by two percent of the balance in your account, and if I will say that the rate has increased by one third of what it was previously, then we can understand each other and agree that both our statements are correct.

The accuracy of an instrument can be stated in several different ways. Assume that an instrument has an indicator range of 50 to 850°C. We may find its accuracy stated as shown in Table 3-2.

The words "of full scale," "of span," "of range," and "of reading" are sometimes omitted through oversight or in the belief that the meaning of "percentage" is clear in context. If doubt exists, the appropriate words should be used to avoid misunderstanding.

High accuracy, meaning low error, is preferred to low accuracy. Whether high accuracy is worth its extra cost and other complications is a question to be answered for each application.

The accuracy of a series of instruments in a loop is generally poorer than the accuracy of each of the individual instruments because the loop accuracy includes the degrading effect of all the individual accuracies. (This is discussed further in Section 10-1.)

- *Accuracy versus Precision.* The words *accuracy* and *precision* do not have the same meaning. *Accuracy* is the closeness by which the apparent value of a variable matches the true value or an accepted standard value of the variable. [12] Or simply, it is a comparison between a reading and the truth.

On the other hand, *precision* is the closeness to each other of successive apparent values of a variable. Or simply, it is a comparison between different readings of the same thing. It is a measure of the scatter of supposedly equal values. Figure 3-4(a) illustrates the difference between *accuracy* and *precision*.

Consider an unadjusted rifle that shoots far to the right of the target but whose shots are very closely spaced. The rifle has low accuracy with high precision, analogous to the illustration of Figure 3-4(a), line (iii).

Table 3-2. Statements of Instrument Accuracy

Accuracy Statement	Remarks
±0.1 percent of full scale (F.S.)	±0.1% × 850°C = ±0.85°C at any point in the range. Acceptable statement. This leads to large errors for readings that are low in the range, thus: For a reading of 500°C, accuracy = ±0.85°C/500°C = ±0.2%. For a reading of 100°C, accuracy = ±0.85°C/100°C = ±0.9%.
±0.1 percent of reading	For a reading of 500°C, accuracy = ±0.1% × 500°C = ±0.50°C. For a reading of 100°C, accuracy = ±0.1% × 100°C = ±0.10°C. Acceptable statements.
±0.1 percent of span	±0.1% × 800°C = ±0.80°C. Acceptable statement.
±0.1 percent of range	Improper statement. A range is defined by two numbers.
±0.1 percent	Incomplete statement. A reference value, such as *full scale*, is required.
±0.85°C over the full range	Acceptable statement.
±0.25°C at 25 percent of span	Acceptable statement.
±0.25°C at 250°C	Acceptable statement.
±0.25°C	Incomplete statement. A reference value, such as *full scale*, is required.

After the rifle sight is adjusted, both the accuracy and the precision are high, as in Figure 3-4(a), line (i).

• *Repeatability versus Reproducibility.* Assume we change an input to an instrument over its full range, by steps up and down, and we measure the corresponding output values. If we do this several times under the same conditions, we end up with the information shown in Figure 3-4(b). Because of various imperfections of the instrument, the several lines of output vs. input in the up direction do not coincide with each other, the several lines in the down direction do not coincide with each other, and the up lines do not coincide with the down lines. If the instrument operated perfectly, there would be only one line instead of six lines.

For a given input, the span of extreme upward values of output is the *upward repeatability*, and the span of downward values is the *downward repeatability*. [75] For that input, the span of the extreme values of up and

down outputs is the *reproducibility*. [76] These terms are illustrated in the figure. The lack of coincidence of lines that would ideally coincide actually reveals the nonrepeatability and the nonreproducibility; nevertheless, these are usually expressed as *repeatability* and *reproducibility*, respectively, in percent of span.

Reproducibility is a more useful word than *repeatability* because it includes the effects of repeatability plus other inaccuracies (drift over a specific period of time, hysteresis, and dead band) for the instrument in question. A reproducibility requirement is not usually put into a specification; it is covered under the blanket term, *accuracy*, which is frequently specified. However, there are applications for which good accuracy of measurement or control are not required but which do require good precision, that is, good reproducibility and predictable performance.

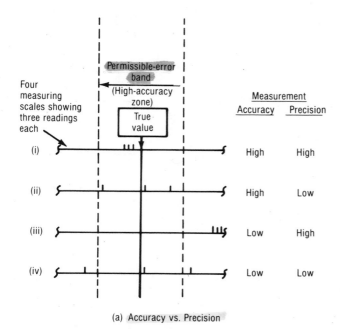

(a) Accuracy vs. Precision

The goal is to measure the true value of a process variable. Any measurement within the permissible-error band is considered to be of high accuracy, outside the band to be of low accuracy. A small scatter of readings shows high precision; a large scatter, low precision.

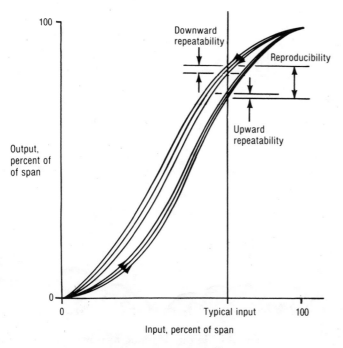

(b) Repeatability vs. Reproducibility

The several curves from repeated trials deviate from each other because of instrument inaccuracy. The ideal would be a single straight line from 0 to 100 percent of the input and output spans.

Figure 3-4. Truth in Measurement

3-2. HOW INFORMATION IS CONVEYED

3-2-1. The Information Chain

Input and *output* are essential words for discussing process instruments.

• *Input* is any information that is supplied to an instrument or other device. The information may come from contact with a process, as in sensing temperature, flow, liquid level, chemical composition, or any other process variable. The information may come in the form of an electric, pneumatic, or other signal from another instrument or a computer that may have modified information it has received from its own input and passes it on. A typical controller modifies its own input and sends it to a control valve.

• *Output* is the information or other effect produced by an instrument or other device. The output may be in one of the following forms:

(1) A signal to one or more other instruments.
(2) An indicator reading.
(3) An on-off light or an alarm bell to show process status.
(4) A record such as a graph, printed chart, or magnetic tape.
(5) A message on a computer video screen.
(6) The process flow manipulated by a control valve. The valve input comes from the controller.

The engineering words *input* and *output* are applied not only to individual instruments but also to instrument systems. (They are even used in non-engineering situations, for example: "We need more input to solve our problem.")

For the instrument loop illustrated in Figure 2-2, Figure 3-5 shows how the output from one instrument is the input to the next instrument. The first input to the loop is information about the process water level. The level sensor, followed by the level transmitter, transform the level into first one signal, then another as the level changes and inform the various receiving instruments about the level. The level indicator, for example, has no output signal but provides a reading of the level for the plant operator. The level controller has a manual adjustment for the desired water level; this is a second input. The controller has an output signal that is the input to the control valve. The control valve is the final control element whose output is the flow of process water to the tank; as the valve input signal changes, so does the water flow change.

Note: Power supplies are not shown.

Figure 3-5. Information Chain for an Instrument Loop

Many, if not most, instruments require a power supply in order to function. The supply is usually electrical or pneumatic. An unmanipulated power supply, as such, is not an input to instruments, even though the power may be indispensable for their functioning; the power supply is simply an energizing medium that contains no information. For example, to watch a television program we turn our set on. This does us no good at all if the selected channel is not broadcasting information.

3-2-2. Types of Instrument Signals

A signal is a variable that is made to represent a different variable in a specific relationship. For example, a process variable, such as liquid level in a tank, may be represented by an air pressure, an electric current or voltage, or a property of some other signal medium. Similarly, a signal may be altered through one or more instruments to provide a desired result, as is done by the level controller in causing the control valve to manipulate the flow of water.

There are three basic types of signals: *binary*, often called *on-off*; *analog*; and *digital* or *pulse*.

● *Binary Signals*. A binary signal is the simplest type of signal. It has two discrete values, ON or OFF, meaning YES or NO, or TRUE or FALSE. The values can be symbolized as 1 or 0, respectively. (*Discrete* means *consisting of unconnected and individual parts.*) An instrument output binary signal changes from one value to the other according to whether the input is less than or greater than some given value. But, depending on the application, the input may be either binary, analog, or digital (see "Analog Signals" and "Digital Signals" below).

An automatic home thermostat is an example of a device providing a binary signal. Assume that the thermostat is set to maintain the room temperature at 70°F but the actual temperature is 65°F. The thermostat is on, an ON signal is being sent to the final control element, which is either a gas valve or an oil pump, and fuel is flowing to the home furnace; the room gets warmer. When the temperature rises to 70 degrees, the thermostat goes off, the signal goes off, and the fuel flow stops. The room cools and the cycle repeats, with the thermostat, in effect, continuously asking: Is the temperature below 70 degrees? If YES, the thermostat commands that the heat be turned on; if NO, it commands that the heat be turned off. The thermostat output signal is always either ON or OFF, with nothing in between.

A *binary signal* is sometimes referred to as a *digital signal* or a *discrete digital signal*. However, a binary signal at any moment states a complete piece of information concerning a condition, whether ON or OFF. The signal state may last only momentarily or indefinitely. It does not have the periodic and repetitious nature of a true digital signal. The output of the level controller in Figure 3-5 may be binary.

Many binary instruments are *direct-acting*, which means that an input that is lower than its desired value causes a low output; an input above its desired value causes a high output. Equally common industrially are *reverse-acting* instruments, for which a low input causes a high output and vice versa. [82]

● *Analog Signals*. You may have seen a jewelry store display of a watch whose face has no dial graduations and which represents a surrender of function to fashion, yet it is readable...somewhat. Despite the lack of markings on the dial, if the watch has the customary 12-hour movement, then we can see the approximate time from the position of the hands. The movement of the hands is analogous to the passage of time because the hands move as the time advances. Hence, such a timepiece is called an *analog-type* watch. An *analog* may be

defined for instrument work as something that moves or changes without a break as something else moves or changes without a break.

Examples of everyday analog instruments are a household thermometer and an automobile speedometer, which give readings that are analogs of the changing temperature and speed, respectively. Such instruments, whose design was not ruled by high fashion, are more practical than the watch and have scale graduations to permit more accurate readings. Instrument signals, also, are often of the analog type, whose value is analogous to the value of the information that went into the instrument. The automatic adjusting of the analog output of an instrument as the input changes is known as *modulation*. [13]

For most analog instruments, the output varies in a one-to-one relationship with the input; for every input value, there is one specific output value. For certain controllers, the output varies as the input varies but also varies with time in a definite way; the output of these controllers can have different values at different times even if the input does not change (see Sections 4-2-3 and 4-2-4).

Input and output ranges are generally stated using percentages that relate to the span of each range (see Section 3-1). However, the scales for a specific instrument are commonly stated in physical terms; for example, a flow input may be 0 to 300 standard cubic feet per minute (scfm) and the signal output may be 4 to 20 milliamperes (mA) dc.

● *Digital Signals*. A digital signal has discrete elements, typically a stream of pulses whose height, frequency, or shape are varied in a way that exactly corresponds to the way that the signal device input changes value. [14] Though the signal has a pulse form, the information it carries may be binary, analog, numerical, or alphabetical. The pulse signal may be converted subsequently into binary or analog form or into a set of discrete symbols such as numerical digits or alphabetical letters to provide an intelligible message for a person. The combined use of letters and numbers is called *alphanumeric*. For example, there are instruments that send out digital signals to represent a flow rate, pressure, or other variable. The ordinary digital watch counts electrical pulses and provides a reading using the ten digits of the decimal numbering system.

Signal converters are available to change digital signals to the other forms or vice versa.

3-2-3. Signal Ranges

Binary signals, by definition, have only two values, a lower one and an upper one. Analog signals have an unbroken series of values from top to bottom of their range. Nominal values of common ranges are shown in Table 3-3. ISA Standards S7.4 and S50.1 define the requirements for pneumatic and electronic analog signals, respectively. [15,16]

In addition to the above signal values, there are other values that may be encountered. There are also electromagnetic types of signal: optical (using light rays), infrared (using heat rays), and radioactive (used for measurements), which are not listed in the table. Also, some measurements are made by acoustic signals.

The reason why some analog signal ranges — for example, 3 to 15 psig or 4 to 20 mA — do not start at zero is to improve the accuracy at the low end of the range. The 3-psig and the 4-mA values are known as *live zeroes*.

3-2-4. Converting Signals

Figure 3-5 shows the output of the level transmitter sending information to six receiving instruments. Assume that the transmitter output signal is compatible with the input requirement for all the receivers except the alarm system and the computer. Figure 3-6 shows how this problem is resolved by adding signal converters. Converters are available for all common combinations of signal ranges and can be developed for all others.

3-2-5. Sharing a Signal Channel

Often a number of signals have to undergo similar modification or be conveyed over a long distance. A system can be designed so that the signals take turns in sharing a single processing instrument or a single communication channel. This type of sharing is known as *timeshare multiplexing*. It requires switching from signal to signal. The switching is termed *scanning*. When performed automatically, the scanning rate may be in the order of 50 points per second or faster. This arrangement requires additional hardware to do the multiplexing but saves money by using fewer signal-processing instruments or less signal wire, or both.

Multiplexing makes data highways possible for use in distributed control systems (see Section 7-3-2). A *data highway* is a single pair of wires that carries information to and from many instruments.

There are a variety of multiplexing arrangements having different levels of sophistication. Figure 3-7 shows the basic concept of multiplexing.

3-2-6. How Instruments Are Identified and Symbolized

Figure 3-6 shows a group of instruments for a single instrument loop. A plant may have many dozens of instrument loops; a nuclear power plant may have up to 40,000 or more individual instruments. An efficient scheme is needed to enable each instrument of the loop to be identified for use in designing the loops, buying the instruments, warehousing, installing, testing, servicing, writing plant operating instructions, and so on. Also, an efficient scheme is required for symbolizing the instruments for engineering and construction drawings. ISA Standards S5.1 [1] and S5.3 [79] address these needs.

Instrument number or *tag number* is the identification number assigned to an instrument. Figure 3-8 shows the ISA method applied to the system shown in Figure 3-6.

Table 3-3. Nominal Signal Ranges

Type	Medium	Values
Binary (on-off)	Electricity: alternating current	0 to 120 volts
	direct current	0 to 24, 48, or 125 volts
	Pneumatic	0 to 25, 35, 100 pounds per square inch gage (psig) (170, 240, 700 kilopascals)
	Hydraulic	0 to 3000 psig (20,000 kilopascals)
Analog (modulating)	Electricity: direct current	–10 to +10 volts, 1 to 5 volts*; 4 to 20 milliamperes*; or 10 to 50 milliamperes
	Pneumatic	3 to 15 psig*; 6 to 30 psig (20 to 100 kilopascals*; 40 to 200 kilopascals)
	Hydraulic	0 to 3000 psig (0 to 20,000 kilopascals)

*The noted ranges are standard. [15,16]

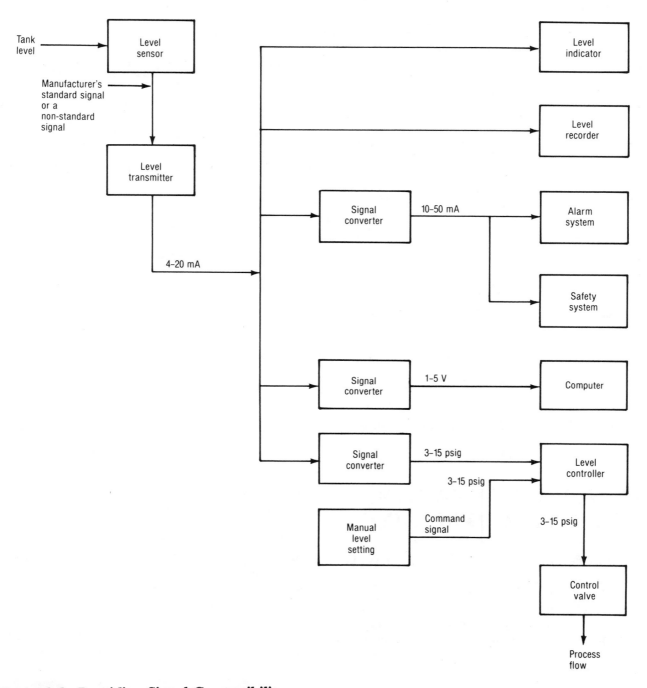

Figure 3-6. Providing Signal Compatibility

The identification numbers used — LV-1, LE-1, and the others — are the individual numbers for the instruments shown.

Scientific Apparatus Makers Association (SAMA) Standard PMC 22.1 [37] has a set of symbols that may be used for diagramming instrument "systems in their basic functional form, exclusive of their operating media or specific equipment detail". ANSI/ISA S5.1 has adopted some of the SAMA symbols.

Where diagrams in this book use instrument symbols with tag numbers, the symbols are taken from ANSI/ISA Standard S5.1-1984, unless stated otherwise.

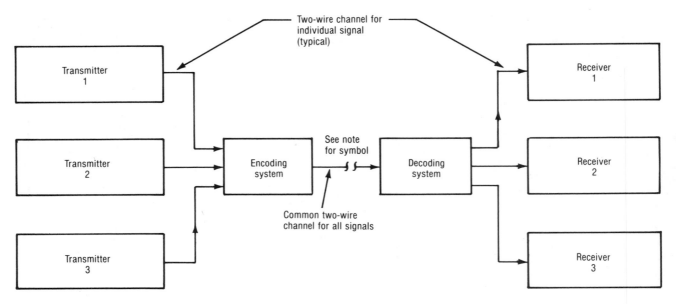

The information from transmitter 1 goes to receiver 1, only. Similarly for 2-to-2, 3-to-3, etc.

The receivers may be instruments or computer points.

The decoding system may be internal to a receiver.

Note: The curved-line symbol denotes an indefinitely long line or distance.

Figure 3-7. The Basic Concept of Multiplexing

3-3. HOW INFORMATION IS DISPLAYED AND RETAINED

Much of the information provided by instruments has to end up in some sort of display that can be read by people. An example is statements of current process conditions that an operator has to keep his eye on in order to see whether the plant is running normally. If the conditions are not normal, then the operator has to decide what to do about it. This is similar to the driver of an automobile checking the dashboard instruments to see that the car is performing properly.

Past information — historical information, as it is called — may be important. The "past" may be five minutes or five months ago. Before the operator decides what corrective action he should take, he may want to look back in time and see how the process behaved: its direction of change —towards normality or away from it, the speed of change, or what caused a failure or accident to happen. Past information is needed also for billing a customer or paying for fuel or raw material. Did pollutant emissions for the past week exceed government

regulations? How efficiently has the plant operated over the past month? How excessive was the operating pressure and for how long?

Though uncommon, even future information can be presented. This makes use of predictive instruments that determine how a process system will behave at some time in the future, based upon what is known about the present state and the characteristics of the system.

Information may be classed as either non-permanent or permanent. The non-permanent information is for immediate use. The permanent information may be for immediate use but is also retained for later use.

3-3-1. Presenting Non-Permanent Information

Non-permanent information may be presented in the following ways.

• *Glasses.* Glasses are the simplest form of display. They consist of a piece of transparent glass or plastic, usually encased in a metal housing, that is connected to the process and shows what is going on inside. Glasses

Symbolism is based on ANSI/ISA Standard S5.1.

The command signal to the controller, LC-1, is omitted here because it is normally built into the controller as an adjustment knob.

The statements in parentheses are for explanation and do not normally appear on a diagram.

Figure 3-8. Standard Symbolism for the Instrument Loop of Figure 3-6

do not have a measuring scale; if they do, they are called *indicators*. The most common glass instrument is the *level glass, level gage,* or *gage glass,* which is connected to a vessel and shows the liquid level (see Section 3-6-2). Another example is the *flow glass* or *sight flow indicator,* which is connected to a pipeline and has a vane, a paddlewheel, or a drip arrangement inside to show flow. Depending on the flow rate, the vane swings open more or less, the paddlewheel turns faster or slower, or the drip stream is stronger or weaker.

• *Indicators.* An indicator has a graduated scale and a pointer or index mark, as typified by a common liquid-in-glass thermometer and most automobile speedometers. The indicator shows the value of a process variable. Examples of analog scales are shown in Figure 3-9. Indicators may be digital and display an exact reading, such as "128.1°F". An automobile speedometer, which shows the speed of travel, is almost always analog. Its companion odometer, which shows the distance traveled, is digital.

(a) Segmental uniform scale

(b)   Vertical uniform scale

(c) Circular uniform scale

Figure 3-9. Typical Analog Indicating Scales

● *Lights.* Small lamps with colored translucent covers are widely used to convey binary information. A lamp is either ON or OFF, sometimes BRIGHT or DIM. The information shown may report that the status is normal, for example, a valve is open, or a pump is running. [52] A light may show that a particular operating sequence is under way or is completed —some automatic dishwashers and clothes dryers have such lamp displays. A lamp that signals which one of a number of normal conditions exists should be called a *pilot light*, though some refer to it as an *indicating light*, or a *monitor light*.

Alarm lights are used to warn of an abnormal situation that may lead or has already led to trouble or danger. [53] Alarms are placed in automobiles for abnormal and potentially hazardous conditions such as low oil pressure, high oil temperature, seat belt not fastened, or door open.

In a process plant, two separate types of alarms are commonly provided. The first alarm is actuated during an abnormal state but before harm has been done. It is a *first-stage alarm* or *preliminary alarm*, which is often compressed to *pre-alarm*. The first-stage alarm notifies the operator, who is then supposed to take action to avoid a potential danger. If and when the condition gets worse, the operator gets a *second-stage alarm* or simply an *alarm*, if a pre-alarm came first. The system that actuated the second-stage alarm may or may not cause an automatic shutdown of a part or all of the process but, in any event, the operator has to go through an appropriate emergency procedure.

Alarm lights and pilot lights are effective only if an operator sees them. They do not necessarily command instant attention. To help the operator, an alarm light may flash to signal trouble. *Flashing* is the rapid alternating of the two binary states of the light, such as ON and OFF. In addition, alarm lights almost always, and pilot lights sometimes, are joined with an audible device — a horn, buzzer, chimes, or even a recorded vocal message that names the process condition — to attract the operator's attention. He then looks for the light that shows what has happened.

Pilot lights and alarm lights in a plant are frequently color-coded to convey more information than merely whether a particular piece of equipment is ON or OFF or whether a condition is normal or abnormal. The color code may show relative importance or priority in case of trouble, designate the process system that is involved, or provide other information.

Pilot lights to denote whether a piece of equipment is operating are often combined with the manual switches that control the equipment. For example, a rotary two-position switch or a double-push button switch can light up a STEAM ON or STEAM OFF light, according to which switch position or button is selected.

 • *Annunciators.* An annunciator is a device or a packaged group of devices that call attention to changes in process conditions that have occurred. An annunciator usually signals abnormal process conditions but may be used also to signal normal process status. It usually calls for interaction with the plant operator.

The attention-getting signals that an annunciator sends out are usually of two kinds, as follows:

(1) *Visible.* An annunciator cabinet usually has a rectangular group of boxes having translucent windows. The number of boxes may vary from one to perhaps 20 wide × 10 high. The windows are translucent, white or color coded. They are engraved with service descriptions, and the engravings are usually blacked in so that they can be read when the windows are lighted from the back. A typical window inscription might be:

SUMP LEVEL HIGH
LAH-215

(LAH-215 is the instrument number for high-level alarm 215. See Section 3-2-6.)

(2) *Audible.* An audible alarm is provided in the control room to supplement the visible alarm, as described above under "Lights". Additional audible alarms may also be placed in other plant locations.

Annunciators have an operating logic, which is a built-in program that controls how visible and audible output signals of the annunciator are turned on by process trouble and how they are turned off. The logic is known as an *operating sequence.* Very many different sequences can be purchased. ISA Standard S18.1 outlines a method for designating sequences but it does not name any sequence as being standard. [54] Two popular sequences are ISA Sequences *A4* and *F1A,* which are described in Figure 3-10.

Annunciators are usually located in a plant control center but are also found in local areas of the plant. Their operating logic may be located in the annunciator cabinet but is often placed in a remote cabinet, especially for large systems.

A so-called *first-out annunciator* is useful where a single process failure may cause a chain of other seeming failures to occur; this results in a group of alarm windows being lighted almost at the same time. What should the operator look at first? The first-out annunciator directs him to the alarm that started the trouble because only that window is flashing; the succeeding, or subsequent, alarm windows have steady lights.

For example, such a progression could happen if there is a loss of cooling water, which causes a failure to cool lubricating oil, which causes a compressor to stop on high oil temperature, which causes a loss of compressor discharge pressure. Several alarms would go on, but the real culprit is the failure of cooling water.

Annunciators may be provided with many optional features in addition to those illustrated in Figure 3-10. Among these features are the following:

(1) *Ringback.* This signals the operator when the process variable for the alarm returns to normal.

(2) *No Lock-In.* This causes the annunciator sequence to return to the normal state without requiring the operator to acknowledge the abnormal state, provided that the process variable is normal.

(3) *Automatic Alarm Silence.* This silences the audible alarm automatically after a fixed time even though the alarm may not have been acknowledged. The visible alarm is not affected.

(4) *Auxiliary Binary Electric Output Signals.* These may be used to operate *slave* alarms or other remote devices. A *slave* alarm copies the alarm action of the main alarm, but usually in a simplified sequence.

Pneumatic annunciators with pneumatic operating logic are available but are used infrequently in the process industries.

 • *Video Screens.* Up-to-date control rooms, and often local operator stations, may have video screens.

ISA SEQUENCE *A* (Formerly *1*)

This sequence is for individual process variables. Each variable has its own visible alarm and shares a common audible alarm with other variables.

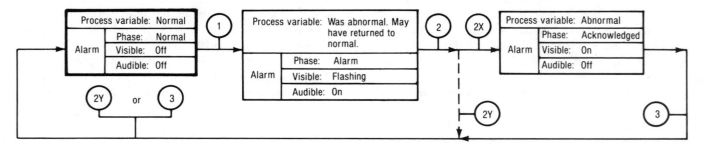

Legend: 1 Process variable has become abnormal momentarily or for a longer time.
2 An *acknowledgment* button is pushed.
2X Path if process variable is still abnormal at time of acknowledgment.
2Y Path if process variable is again normal at time of acknowledgment.
3 Process variable is again normal.

(a) Annunciation for a single process variable

ISA SEQUENCE F1A (Formerly *4A*)

This sequence is for a group of process variables. Just as for sequence *A*, above, there is an individual alarm window for each process variable and a common audible alarm for the group of variables. However, the visible display for the group's first alarm to be actuated is different from the displays for subsequent alarms of the group. Thus, the initial trouble of a progression of troubles can be identified.

● For the first abnormal process variable (first out) of the group, the alarm sequence is the same as for sequence *A*. Any process variable of the group may be the first out.

● For a subsequent abnormal process variable, the typical sequence is:

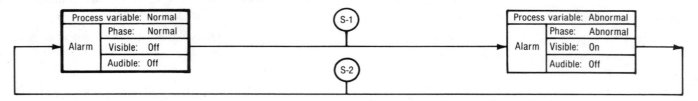

Legend: S-1   Process variable is abnormal. (Acknowledgment is not required.)

S-2   Process variable is again normal.

(b) First-out annunciation for a group of process variables

Figure 3-10 Two Popular Annunciator Sequences (based on ANSI/ISA S18.1-1979 [R1985])

The screens can display any kind of information that can be put on paper: drawings, lists, instructions, etc. Video may be found with computer control and distributed control (see Section 7-3-2), and programmable logic control (see Section 6-4-2).

3-3-2. *Presenting and Retaining Permanent Information*

● *Chart-Type Recorders.* The most common type of process recorder is essentially a machine that draws graphs on a paper known as a *chart*. The most common type of graph shows how a measured variable, such as flow, varies with time. This uses a clock drive to move the chart while an actuator that responds to a flow signal moves a pen across the chart. The chart provides analog information for current use by the operator, and it may be filed for future reference.

Charts in a control room are usually straight, usually 4 inches or 12 inches wide, and in long lengths that are typically adequate for a month's continuous use; these are termed *strip charts*. Strip charts may be used in the field, but charts for field use are usually round, 12 inches in diameter, and are usually good for 24 hours of use, sometimes for seven days or a month.

Recorders may be either single-point type or multi-point type, depending on the number of measured variables that are to be recorded. A single-point instrument serves one variable, such as pressure or temperature. A multipoint instrument serves more than one variable, such as two or more pressures or flows, or one pressure and three temperatures.

The single-point recorder has a single pen. The multipoint recorder may have either one or many pens. The points of a multipoint recorder may share the same chart space and their lines may cross each other; each of the various records is identified and distinguished from its companions by a color code or printed numbers. On a 12-inch strip chart there is usually only one pen, which switches among 24 input signals. (This is an example of a multiplexing device, as described in Section 3-2-5.) Some charts carry eight or more parallel pens in separate channels without overlap.

Chart recorders are available with alphanumeric readings and with auxiliary output signals for switching or controlling.

Data loggers are often used to keep track of all the important variables in the plant. They record operating data, perform calculations, and check for alarm conditions. They record routine operating data by switching internally from point to point; however, they give recording priority to abnormal conditions and then switch to high-speed printing.

The record they provide is an alphanumeric tabulation, which contains data such as the date and time, point identification, service, and value of the process variable. The printing is normally black but may be switched to red in order to highlight abnormal conditions.

Event recorders perform some of the same functions as data loggers perform. They may normally be kept in a standby state, ready to start up immediately upon the occurrence of an abnormal condition and record the abnormalities. Their printing speed is very high, with times reported down to a hundredth of a second. This makes it possible to see what conditions initiated a sequence of alarms (see the above discussion of first-out annunciators and Figure 3-10).

• *Printers.* Printers are furnished as integral parts of certain recording systems and are also furnished routinely as so-called *peripheral devices*, auxiliaries for computers and distributed control systems. They may also be used to draw graphs of data stored in a computer or distributed control system.

• *Other Recording Devices.* A variety of other means are available for storing operating information. These include the following:

(1) *Magnetic tape.* The information may be analog or digital. It cannot be read directly.

(2) *Video tape.* This is similar to magnetic tape but can also store pictures and sounds.

(3) *Video copier.* This can photographically copy whatever a video screen displays.

(4) *Computer disks.* These are the disks that are normally used with the widespread computer. They include the inexpensive so-called *floppy diskette* in the 5-1/4-inch and 8-inch sizes, the large-memory *hard disk*, and the 3-1/2-inch disk.

3-4. HOW PRESSURE SENSORS SENSE

3-4-1. Units of Pressure and Force

Pressure and *force* are very important factors for industrial processes and the world we live in. It may be helpful to review the difference between these two words.

If we place a 10-inch cube of metal weighing 260 pounds on a table top, the cube imposes a downward force of 260 pounds on the table. The contact area of the cube with the top is 10 inches by 10 inches, or 100 square inches. The pressure, however, is the force divided by the contact area, 260 pounds divided by 100 square inches, or 26 pounds per square inch. This downward pressure is not on the whole top but only on the 10-inch by 10-inch portion of the top. If the top were 4 feet square, 2304 square inches, and were covered solidly with 10-inch-high pieces of metal, the downward force on the table would be 60,000 pounds, or 30 tons. The table would have to be very special not to collapse under that load, but the pressure on the table top would still be only 26 pounds per square inch (60,000 pounds divided by 2304 square inches).

If a person blows air through his lips, he creates a small, low-pressure wind that can deflect a sheet of paper and not much more. An atmospheric wind with the same low pressure can move sailing ships because the pressure is exerted over the large area of sail, thereby creating a large force.

Figure 3-11 shows two vertical pipes on supports, one pipe having an inside diameter of 2.0 inches, the other an inside diameter of 8.0 inches. The bottom of each pipe is closed, the top open. Both pipes are filled with cold water to a level of 27.68 inches. For each pipe, the downward force on its support is the same as its weight of water (ignoring the weight of the pipe). The force is

Figure 3-11. Force and Pressure at the Bottom of Two Water Tanks

The outlet pressure of the blower is 1½ inches WC (water column), gage

Figure 3-12. Measuring Low Pressure

3.14 pounds for the 2-inch pipe and 50.28 pounds for the 8-inch pipe. But the pressures at the bottom of both pipes are identical at one pound per square inch; that is because the pressure depends on the height of water, not the volume. The height of liquid is known as *head*.

The head of atmospheric air on the Earth extends to the stratosphere and beyond. The weight of the atmosphere and its force on the entire surface of the Earth are colossal but, on one square inch of soil at sea level, the force is approximately 14.70 pounds; therefore, the pressure is 14.70 pounds per square inch.

In the United States, we use various units to express pressure in engineering work. Most used is *pounds per square inch*, symbolized as *psi*. For very low gas pressures, we use *inches of water*, symbolized as either in. WC, meaning *water column*, or in. H₂O, the chemical symbol for water. Measured inches of water is the height or head of water whose bottom pressure equals the pressure of the process variable sensed by the instrument (see Figure 3-12). One inch of cold water equals 0.0361 pound per square inch.

Pump pressures are frequently expressed as *feet of head*, shown as *ft head*. This refers to a theoretical column of the pumped liquid that the pump pressure could support. The height of the theoretical column, the pump head, changes as the density of the liquid increases or decreases. *Density* is the ratio of the mass of a substance to its volume. *Mass* is a basic measure of the quantity of matter.

Thus, oil floats on water while stones sink because of their different densities. The density effect is often significant when head is the sensed variable used for making measurements of flow, level, or other variable.

Barometric pressure is frequently expressed in *inches of mercury, (in. Hg)*, or as *millimeters of mercury (mm Hg)*. Here we refer to a column of liquid mercury. Measured at sea level, barometric pressure is normally 29.92 in. Hg, equal to 760.0 mm Hg and 14.70 psi. This is the so-called *normal barometer*, whose frequent changes affect all of us and gage-pressure instruments (see Section 3-4-2).

For force, the United States uses the engineering unit *pound*, meaning *pound avoirdupois* (lb), which has 16 ounces to the pound. *Ton* is also used, meaning the *short ton*, which has 2000 pounds.

The United States units of force and of weight are the same except that an object weighing 10 lb at sea level exerts a downward force of 10 lb but, when raised to a higher elevation, the force is less than 10 lb. This is because the weight of an object is really the downward pull on the object caused by gravity; the farther the object gets from the center of the Earth, the less is the downward force and the corresponding weight. What remains unchanged about the object is its mass. However, we are seldom concerned about mass when we deal with process measurement and control. We usually deal with weight. However, there are cases where gravity effects may be significant (see Section 8-2-6).

A helter-skelter system of measuring units used to exist throughout the world and still exists in the English system that we largely continue to use in the United States. In 1791, the French Academy of Sciences established the metric system, a new, rational system of units using decimal relationships. The metric system spread throughout the world. In 1960, 36 countries, including the United States, modified the metric system and established the International System of Units, which is internationally symbolized as *SI*. The SI system is becoming the world standard, although United States practice does not make general use of SI or of its predecessor, the metric system. [17]

The SI unit of pressure is the *pascal* (Pa). There are variants, such as *kilopascal* (kPa, a thousand pascals), *millipascal* (mPa, a thousandth of a pascal), etc. A pascal equals one newton per square meter; a kilopascal equals 0.1450 psi. The *newton* (N) is the unit of force, independent of gravity, and equals 0.2248 pound of force. The *meter* (m) (sometimes spelled *metre*) is the unit of length and equals 39.37 inches.

The metric unit of force is the *gram*, which is being replaced by the newton. The metric unit of pressure is *kilograms per square centimeter*, which is being replaced by the pascal.

3-4-2. Kinds of Pressure

Figure 3-12 shows pressure being measured for process air whose pressure is higher than atmospheric pressure. The higher pressure pushes the water down on one side of the measuring instrument while the water is pushed up against atmospheric pressure on the other side; the difference in water levels is 1-1/2 inches. The pressure measurement is made using atmospheric pressure as a reference. Whenever the reference pressure is atmospheric, the measured pressure is called *gage pressure*. (*Gage* is frequently spelled *gauge*.) Therefore, the pressure on the discharge side of the blower is 1-1/2 inches of water gage.

The gage pressure may be either positive (that is, above atmospheric as in the figure) or negative (below atmospheric). A negative gage pressure is a vacuum, which, if it existed in the process, would suck the water of the instrument higher on the left-hand side, lower on the right-hand side.

If all the air were removed from the tubing on the atmospheric side — the right-hand side — of the instrument and if the open end were sealed tightly, then the reference pressure would be absolute zero instead of atmospheric. This would give us an *absolute-pressure* measurement, which is always positive.

The instrument in Figure 3-12 is connected to the discharge side of the air blower. If we connect the open end of the instrument to the suction side of the blower,

then the instrument will sense the difference in pressure, the *differential pressure* or *pressure differential*. In writing, these terms may be symbolized by ΔP or delta P. The instrument would show how much the blower increases the pressure of the air in going from blower suction to discharge. The blower is powered by an electric motor or some other type of drive and imparts energy to the air passing through the blower. This energy is a process input.

The kind of instrument we have illustrated to measure blower pressure is a *U-tube manometer*. Its various uses are shown in Figure 3-13. Note the basic similarity of cases (a), (b), (c), and (d); in all four cases, the right-hand side of each manometer is connected to a process pressure. The essential difference among the cases is the nature of the reference pressure on the left-hand side: The left-hand side is connected to the atmosphere in cases (a) and (b), to a total vacuum in case (c), and to a second process point in case (d). In cases (a) and (b), an increase in atmospheric pressure pushes the liquid down on the left side of the manometer so the liquid on the right side moves up, thereby reducing the reading in (a) and increasing it in (b); this reflects the fact that a given physical pressure of the process gives a gage-pressure reading that depends on the atmospheric pressure, which is changeable. The relationship between absolute and gage pressures is shown in Fig. 3-13(e).

The word *pressure* is sometimes ambiguous. It is used in a generic sense to denote any of the following: positive gage pressure; negative gage pressure, which is a vacuum; absolute pressure; or differential pressure. It is also used specifically to denote a positive gage pressure as opposed to a vacuum. Thus, we have a pressure regulator that prevents a tank from both bursting from pressure and collapsing from vacuum. If we are told that a pressure is 32 psi, for example, then we cannot be sure whether the measurement is of gage pressure, absolute pressure, or differential pressure. Symbols can be used to avoid ambiguity, as follows:

Gage pressure	psig
Absolute pressure	psia
Differential pressure	psi or psid

Vacuums and pressures stated in terms of head of liquid are usually understood to be gage pressures, but if there may be misunderstanding, then a clarifying word or symbol may be helpful.

3-4-3. Types of Pressure Sensors

The basic design of several common types of pressure sensors is described below. Some of these instruments may indicate the measured pressure. They are often

Figure 3-13. Types of Pressure Measurement and Pressure Relationships

used for other functions, including transmission, switching, controlling, and recording.

• *Manometer.* The principles of manometers for measuring pressure have been covered in Section 3-4-2. The glass U-tube manometer is used for pressures and vacuums that are close to atmospheric pressure. Metal manometers are used for higher pressures. The manometer is important as one of the types of highly accurate measuring standards that are used for calibrating plant instruments (see Section 3-1-4).

• *Bourdon.* The most widely used pressure instrument in a plant is usually the pressure gage, which is used for local measurements. The most common sensing element among all types of industrial pressure instruments is the Bourdon tube, which is illustrated in Figure 3-14.

The Bourdon tube was patented by Eugene Bourdon of France in 1840. The tube is of metal, is curved, has an oval cross section, and acts as a spring. The inside of the tube is at process pressure, and the tube is inside a case that is generally at atmospheric pressure. Thus the tube senses the difference between the pressures that are inside and outside the tube; therefore, the difference is the gage pressure. As the process pressure increases, the cross section tends to become less oval, more circular, and the tube tends to straighten out. This causes the far end of the tube to move. The movement, through links

(a) For gage pressure

(b) For absolute pressure

(c) For differential pressure

Figure 3-15. Basic Design of Bellows Sensor

Figure 3-14. Basic Design of Bourdon Sensor

and gearing, causes a pointer to rotate around a circular graduated dial. There are variations on the basic design.

A round-dial pressure indicator used to measure tire pressure is apt to have this type of construction.

- *Bellows.* Another common type of sensor uses a metal bellows or capsule that performs somewhat as the Bourdon sensor does. Pressure contracts the bellows, release of pressure permits the bellows to relax. Movement of the bellows is communicated to a pointer, which permits us to read the pressure.

Figure 3-15 shows how a bellows-type pressure instrument is modified to measure absolute pressure and differential pressure. Other types of pressure sensors can be modified similarly to measure any type of pressure, assuming that the sensitivity and structural strength of the sensor fit the application.

- *Diaphragm.* The force from the process pressure acting against a flexible diaphragm is opposed by a spring. Increasing pressure results in movement of a pointer. This is illustrated in Figure 3-16.

- *Strain Gage.* When a solid object is subjected to pressure, it deforms and is said to undergo *strain*. A strain gage sensor, or simply *strain gage*, is an electronic device with an electrical capacitance or resistance that changes as the sensor undergoes strain. When the sensor feels a process pressure, the sensor's capacitance or resistance can be measured and correlated with pressure.

Figure 3-16. Basic Design of Diaphram Sensor

3-5. HOW FLOW SENSORS SENSE

3-5-1. Units of Flow

In the United States at present, the English system of units is in general use for plant measurement of process flows. The units depend on the physical state of the flowing substance — gas, vapor, liquid, or solid (particles). For flow rates, the units are commonly as shown in Table 3-4. Note that some of the measuring units are based on volume, some on weight. However, all flow measurements may be based on weight. The unit *gallon* in the table refers to the U.S. gallon, not the British imperial or the Canadian gallon.

The word *flow* is ambiguous. It usually means *flow rate* but, loosely used, it sometimes means *volume*, as in "stop the pump after a flow of 1000 gallons".

The SI system uses the following units for flow rates:

- Volume-based: Cubic meters per second (m^3/s). (1 cubic meter = 264.2 gallons. The volume unit *liter* —sometimes spelled *litre* — is recognized by SI but its use is discouraged; *cubic decimeter*, which is equal to *liter* for practical purposes, is preferred. 1 cubic decimeter = 0.2642 gallons)

- Mass-based: Kilograms per second (kg/s) (1 kilogram = 2.205 pounds)

The volume-based units for gas flow, such as cfm, have a significant refinement, resulting in units of scfm or acfm, meaning *standard cubic feet per minute* or *actual cubic feet per minute*, respectively. The actual volume of a given gas flow depends on the existing temperature and pressure of the gas. The actual volume is converted to a standard volume by calculating what the volume would be if the gas were at a standard temperature and a standard pressure. Thus, a meter that measures the flow of fuel gas to the home senses acfm, but the monthly bill we receive is based on that measurement corrected to state the scfm. We pay for the correct amount of gas we consume, neither more nor less.

The so-called standard conditions for gas measurement, which are referred to as *base conditions*, must be defined for each contract or project because there are no standard "standard" conditions. Base pressures that are used include 14.4, 14.65, 14.70, and 14.9 psia. Base temperatures include 32, 60, 68, and 70°F. The mass or weight of a given type of gas in a "standard" volume depends on which base conditions are used. The differences among the base conditions and the distinction between acfm and scfm can be important for a plant that

Table 3-4. Flow Measuring Units

Fluid State	Units
Gas (see Appendix B)	Cubic feet per hour (cfh), cubic feet per minute (cfm)
Steam and other vapors (see Appendix B)	Pounds per hour (pph)
Liquid	Gallons per minute (gpm), barrels per day (bpd) in the petroleum producing and refining industries (1 barrel = 42 gallons), cubic feet per second (cfs) in the water and sewage industries
Solid (powders or lumps)	Pounds per hour (pph), tons per hour (tph)
Slurry (liquid-and-solid mixtures)	Pounds per hour (pph), tons per hour (tph)

buys or sells any gas, purchases equipment such as a gas compressor, or checks on plant performance.

An analogous situation regarding standard conditions exists for the volume-based measurement of liquid flow. There are *standard gallons* and *flowing gallons*. However, temperature is the major influence on liquid volume; the effect of pressure is usually unimportant. In the United States, the most commonly used base temperature for liquids is 60°F. Gasoline stations measure the volume of gasoline delivered to a car at the flowing temperature but the measurement is automatically corrected to a standard temperature, and we pay accordingly.

Mass-based flow measurements are not sensitive to temperature or pressure and are, therefore, not subject to misunderstanding. The same is true, practically speaking, for weight-based measurements.

3-5-2. Types of Flow Sensors

There are a bewildering number of kinds of sensors to measure the fluid flow rate and volume. Many sensors can be grouped into several different types, as described below.

● *Differential-Pressure.* This type of flow sensor is the one most commonly used in industrial work. The most common of the type is the *orifice plate*, or *thin-plate orifice*, which is usually a circular, thin, metal plate having a concentric hole with a sharp inlet edge. [18] The plate is set into the pipeline.

Wherever there is an orifice plate — or any other restriction or obstacle to straight flow — the fluid pressure downstream of the plate is lower than the upstream pressure. This difference in pressure created by the plate is called *differential pressure*, which has a value of zero at zero flow and which increases as the flow increases. An

instrument, such as an indicator or transmitter, that can sense the differential pressure can then be a flow meter. The orifice plate is a *primary element* and the indicator is used as a *secondary element*; together they measure the flow rate.

Figure 3-17 shows the orifice plate installed in the pipeline and the resulting streamlines, which are the paths followed by fluid particles. The streamlines bend with the pressure changes in the pipe, and the figure shows how the secondary instrument is connected to sense differential pressure, thereby measuring the flow.

Figure 3-17. Using an Orifice Plate to Measure Flow Rate

The figure also shows a partial recovery of pressure, leaving a permanent loss of pressure from the fluid constriction, just as a kitchen faucet that is only partially open restricts the flow and causes the water to come out as a weak stream; this permanent loss is called *pressure loss.*

Mathematical equations are used to size the hole accurately according to the fluid, the maximum flow rate, the flow conditions, and the exact locations of the differential pressure-sensing connections on the pipe. There are several standard pairs of locations for the upstream and downstream connections, which are called *taps.* The most common arrangement uses *flange taps,* for which the center of the connections is 1 inch before and 1 inch after the respective faces of the plate. Other arrangements are known as *radius taps, vena contracta taps, pipe taps,* and *corner taps.*

The plate depicted in Figure 3-17 may have variations: greater thickness, off-center location for the hole, noncircular hole, curved instead of sharp-edged inlet for the hole, different metals for the plate — all for specialized purposes. Sometimes a specially shaped tube — known as a Venturi tube, named for the 18th–19th century physicist G. B. Venturi, or one of its variants — is used instead of an orifice plate. The tube works on the same principle as the plate does.

For a differential pressure type of flow element to measure with maximum accuracy, the flow must have a *normal velocity profile,* or a *fully developed velocity profile,* which is a symmetrical velocity pattern of the fluid all around the pipe. If a run of pipe is straight, without internal disturbance, and of sufficient length, then the velocity profile is normal. Otherwise the profile is distorted, and the flow measurement is less accurate. Recommended minimum lengths of straight pipe to provide a normal profile are established by the American Society of Mechanical Engineers and the American Gas Association. [19, 72]

In order to meet the accuracy requirement for flow measurement, a pipe run may have to be lengthened. This would increase the cost and may be physically difficult to do. An expedient sometimes used to shorten the straight-run requirement is to install straightening vanes, which are a short length of a stack of small pipes, or the equivalent, inserted inside the main pipe.

Another widely used differential pressure type of flow element for general process use is the averaging-type Pitot tube, which is a modern modification of the traditional Pitot tube, invented by the Frenchman, Henri Pitot, in the 19th century. Essentially, the Pitot tube now uses two small tubes that enter the side of the main pipe and extend across it. One tube has holes facing upstream, the other has one hole facing downstream, so the tubes have a differential pressure that depends on the flow

rate. A secondary instrument connects to the tubes and is the flow meter.

For many kinds of instruments, the output varies proportionally with the input. A percentage change in one is matched by the same percentage change in the other. This is generally true for pressure indicators, for which doubling or tripling the input doubles or triples the reading. This relationship is called *linear* or *uniform.*

The same is true for many other instruments, including many flow instruments. However, it is not true for differential pressure flow instruments, for which the output varies as the square of the flow. For example, if an orifice plate gives an output differential pressure of 20 inches WC for a 300-gpm flow, then doubling the flow to 600 gpm quadruples the output to 80 inches WC. Tripling the flow causes the output to be nine times as large. This is an example of a *nonlinear* or *non-uniform* relationship. Unless this nonlinearity is altered by a special instrument, it requires a non-uniform measuring scale. (For more on measurement nonlinearity, see Section 4-2-5(11).)

• *Rotameter.* A rotameter, also known as a *variable area meter,* has a vertical tapered tube, small end down, with fluid flow upward. Inside the tube is a weight that rests at the bottom of the tube but that is forced higher and remains suspended as the flow increases. The weight is therefore called a *float.* It is sometimes called a "rotor" even though it does not necessarily rotate.

The tube may be of glass with graduations to permit direct reading of the flow rate, but the tube is often made of metal for strength. The float may have a vertical stem attached on top to permit connecting to secondary functions, such as transmission (see Figure 3-18). The rotameter scale varies proportionally with flow, and is therefore linear.

• *Other.* Other types of flow sensors are used in process plants, and each type has its own set of pros and cons for different applications. Included among the other sensors are the following:

(1) Positive displacement, like the water or gas meter used in homes.
(2) Magnetic.
(3) Turbine.
(4) Vortex.
(5) Thermal.
(6) Ultrasonic.
(7) Weir and flume, to measure flow in open channels.
(8) Weight (for flow rate of solids on a conveyor).

The positive-displacement and turbine sensors rotate and can be supplied with a transmitting element that

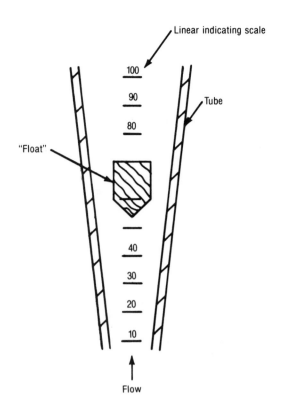

Figure 3-18. Basic Design of Rotameter

sends out a pulse with every rotation of the sensor. This output signal is digital and can be used directly or be converted to other forms.

3-6. HOW LEVEL SENSORS SENSE

3-6-1. Units of Level

Level measurements are usually expressed as a distance or as a percentage of level span, but are sometimes stated as a volume or weight. For height, the units may be inches or feet, whichever is more convenient. The SI unit for height is generally the meter.

But what shall be the datum, the starting point for the measurement? Consider the tank of Figure 3-19, which is a composite of different situations. The level-measuring instrument has a range whose lower and upper limits

may be chosen differently according to the situation and differently by different engineers. Real possibilities are the following.

- For the Lower Limit of Measurement
    Bottom of normal band
    Bottom of overall operating band
    Centerline of side outlet connection
    Inside bottom of side outlet connection
    Lower tangent line of tank
    Extreme bottom of inside of tank

- For the Upper Limit of Measurement
    Extreme top of inside of tank
    Upper tangent line of tank
    Centerline of overflow connection
    Inside bottom of overflow connection
    Centerline of side inlet connection
    Top of overall operating band
    Top of normal band

The combination of limits that is chosen depends on their significance to the process and to the plant operators. The limits should be clearly defined to avoid uncertainty or misunderstanding as to what a given level reading really means.

Whatever the measuring range may be, the upper sensing point must be at the same elevation as or higher than the upper limit of the range, and the entire lower sensing line must be at the same elevation as or lower than the lower limit of the range. Otherwise some of the range will not be useable.

Let us choose a measuring range from the bottom of the side outlet to the bottom of the overflow connection, and say the span is eight feet. A level indicator could be calibrated either 0 to 100 percent or 0 to 8 feet, using the lower limit of the range as the reference level or datum; or –50 to +50 percent or –4 to +4 feet, using the midpoint of the range as the datum; or 353 to 361 feet, using the elevation of the plant as a datum. All these ranges are equivalent but the most meaningful one for the application should be selected.

A level instrument may be calibrated in terms of volume, such as gallons or barrels, or in terms of weight, generally as pounds. These calibrations are simple for a cylindrical vertical tank because liquid volume and weight vary uniformly with level. If the same tank were placed horizontally, the relationship of volume and weight to level would be very nonlinear because of the curvature of the tank.

In the SI system, the volume units for level may be cubic meters, or cubic decimeters, and the weight units would be newtons (comparable to the US term, pounds of force).

Figure 3-19. Possible Limits of Level-Measurement Ranges

3-6-2. *Types of Level Sensors*

There are many types of level-sensing instruments, among which are the following.

● *Float.* One of the major types of level sensors uses a *float* on the surface of the liquid in a tank, rising and falling with the level. The device used to refill a toilet tank in the home has a float sensor. Figure 3-20(a) shows examples in which a float operates (a) an electric switch when the level is above or below some fixed value, and (b) a level indicator. The plant float action can provide level indication, transmission, etc. The float may be mounted inside the tank but, more often, it is placed inside a metal chamber that is piped to the tank and fills with liquid to nominally the same height as exists in the tank.

● *Displacement.* Displacement-type level sensors use a vertical cylinder called a *displacer*, which is sometimes erroneously called a *float*. While a float stays on the liquid surface, a displacer always tends to sink to the bottom of the tank because it is heavier than the liquid it displaces. Figure 3-20(b) illustrates how the weight of the displacer is counterbalanced by a spring. The displacer, like the float, may be mounted inside or outside the tank.

When the level is at the bottom of the range, the spring bears the entire weight of the displacer and the indicator shows zero level. A transmitter output signal would be at the bottom of its range. As the level rises, the displacer becomes partially submerged, and a buoyant force reduces the apparent weight of the displacer by the weight of liquid that is displaced by the displacer. When the level reaches the top of the range, the displacer is fully submerged, the apparent weight of the displacer is minimum, the indicator shows 100 percent level, and the transmitter output signal would be at the top of its range.

The buoyant force follows Archimedes' Principle, which was first stated by Archimedes of Greece in the 3rd century B.C.

● *Head.* The head of liquid in a tank is equivalent to a pressure. By measuring the pressure, we can know the liquid level. The pressure is measured using the pressure instruments described in Section 3-4 and shown in Figure 3-20(c). (The convention used for identifying such instruments is to label them as level instruments, according to their use, not as pressure instruments, according to their construction. [60])

The figure shows an open tank, which is at atmospheric pressure. The level instrument to measure the liquid head can be an ordinary pressure indicator or other instrument that measures gage pressure. However, this indicator cannot be used for the closed and pressurized tank of the figure because the reading could be seriously false. The indicator would sense not only the head of liquid but also the pressure above the liquid inside the tank and would indicate the sum of the pressures.

To measure only the head, we need to use a differential pressure instrument. We connect one side of this instrument to sense the liquid head plus tank pressure, the other side to sense only the tank pressure. The instrument subtracts the tank pressure from the sum of head plus tank pressure, and the resultant is pure liquid head. If the tank level is the same in the tanks of Fig. 3-20(a), both their instruments will give the same reading.

● *Glass.* A basic level instrument that is widely used, very often in conjunction with other types of level instrument, is a *level glass*, *level gage*, or *gage glass*. This is a vertical length of either glass tubing or a metal chamber having one or two glass windows. The tubing or chamber is piped to the tank and shows nominally the same liquid level as exists in the tank.

● *Other.* Other types of level sensors include:

(1) Capacitance probe. A *probe* is a slender rod that is inserted into the process fluid.

(2) Conductance probe.

(3) Radiation, using a nuclear source.

(4) Ultrasonic.

(5) Optical.

(6) Thermal.

(7) Rotating paddle, used to measure solids level in a bin.

(8) Weight. A weight instrument senses the weight of the tank plus its contents but can generally be adjusted to subtract the tank weight automatically. The resultant is the weight of the contents.

3-7. HOW TEMPERATURE SENSORS SENSE

3-7-1. *Units of Temperature*

In the United States, the temperature in process systems is usually expressed in *degrees Fahrenheit*, which is symbolized °F and is named after Gabriel D. Fahrenheit, a German physicist of the 17th and 18th centuries. The Fahrenheit scale is based on the freezing point of water, 32°F, and the boiling point of water, 212°F at normal atmospheric pressure. The lowest possible temperature, absolute zero, is at –459.69°F. An absolute temperature scale, in *degrees Rankine* (°R), has degrees Rankine equal to degrees Fahrenheit plus 459.69. Thus, a room temperature of 72°F equals 531.69°R. The unit "degrees Rankine" is named for the 19th century Scottish physicist William J. M. Rankine.

A float-actuated level switch

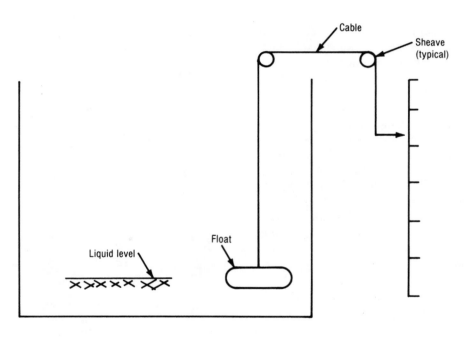

A float-actuated level indicator

(a) Basic designs of float-type instruments measuring liquid level

Figure 3-20. Basic Methods of Measuring Liquid Level

(b) Basic design of displacement-type level sensor

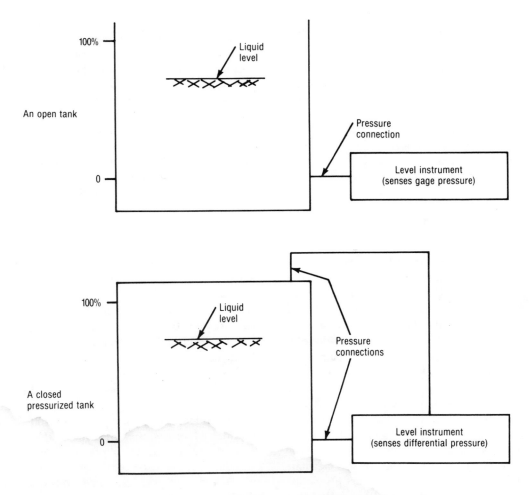

(c) Basic methods for measuring level by sensing liquid head

Figure 3-20 (continued)

In the SI system, the usual temperature scale is in *degrees Celsius*, named after Anders Celsius, the Swedish inventor of the centigrade scale in the 18th century. What used to be *degrees centigrade* should now be called *degrees Celsius*, symbolized °C. This scale defines the freezing point of water as 0°C and the boiling point as 100°C. The corresponding absolute temperature scale, in *kelvins*, begins at absolute zero, or –273.15°C. Thus, 20°C equals 293.15 K. The kelvin is named for the 18th–19th century Irish-British physicist William T. Kelvin.

One degree Celsius equals 1.8°F. The number of degrees Celsius equals the difference, degrees F minus 32, divided by 1.8. –40°F equals –40°C.

3-7-2. *Types of Temperature Sensors*

Most temperature sensors that are inserted into a process fluid are provided with a protective enclosure (see "Temperature Wells" in Chapter 9).

- *Bimetallic.* The most common type of sensor for reading a process temperature locally is a bimetallic thermometer (Figure 3-21 (a)). This sensor uses two thin strips of different metals bonded together over their length. When the bonded metals are heated, they expand unequally and the strips bend toward the side that expands less. The bending causes movement of the end of the bimetal, which results in a pointer moving to indicate temperature.

The same bimetallic principle is used to actuate electric switches and is found in electric irons and thermostats in the home.

- *Thermocouple.* A thermocouple consists of a pair of wires, each wire being of a different metal. When the wires are joined at one end and that end is heated, a small electrical voltage can be measured at the open end of the pair. (*Voltage* is an electrical force, usually stated in the unit of *volt* (V), or *millivolt* (mV, one thousandth of a volt) for thermocouples. Voltage is the force that pushes electrons (subatomic electrical particles) through an electrical circuit. The flow of particles is a *current*, usually measured in the unit *ampere* (A), or *milliampere* (mA). The material property that impedes the flow of current is *resistance*, usually measured in the unit *ohm* (Ω)). The volt is named for the Italian Alessandro Volta; the ampere for Andre M. Ampere of France; the ohm for the German Georg Simon Ohm; all were 18th–19th century physicists.

The joined end of the thermocouple is called the *measuring junction*, the open end is the *reference junction*. The voltage increases as the heated end gets hotter, so measuring the voltage of a thermocouple submerged in a process fluid is equivalent to measuring the temperature of the fluid. The basic measuring system, using a millivoltmeter to measure the voltage, is shown in Figure 3-21(b). The thermocouple response is the Seebeck effect, named after Thomas J. Seebeck, a German physicist who discovered it in the 19th century.

Standards have been established for the type and composition of the metals and alloys that are used for thermocouples. [20] A commonly used type of thermocouple is *iron-constantan* (ISA type J), which has one wire of pure iron and another of constantan. Constantan is a specific alloy of nickel and copper.

The measuring junction and reference junction are sometimes referred to as *hot* and *cold junctions*, respectively, but this usage is not recommended. The words *hot* and *cold* can be misleading because thermocouples can be used to measure very low process temperatures, in which case the "hot" junction is colder than the "cold" junction.

- *Resistance.* Another important class of temperature sensors is the resistance type. This type depends on the fact that the electrical resistance of a wire increases as the wire temperature increases, as first described by the German-British inventor, Sir William Siemens, in the 19th century. If the electrical resistance of a given type of wire submerged in a process fluid is measured, that is equivalent to measuring the temperature of the fluid.

The sensors may be made of various metals, in which case they are known as *resistance temperature detectors* (RTDs) or *resistance thermometers*. Pure copper, nickel, and platinum are widely used for RTDs, and a metal alloy is also used. The sensors may also be made of metal oxides, and are then known as *thermistors* (THERMal plus resISTOR).

- *Filled Thermal System.* Filled-system temperature sensors have a small metal bulb that is connected to a pressure sensor by means of small-bore tubing. The whole system is filled with a fluid and is then hermetically sealed. If the bulb is submerged in a hot or cold process fluid, the fluid in the bulb gets hot or cold and the pressure in the filled system rises or falls. The pressure sensor responds to the change in pressure by, for example, moving a pointer to indicate the new temperature (see Figure 3-21(c)).

The filled system may be filled 100 percent with a liquid, such as mercury, or a gas, both of which expand as they are heated. Or it may be partially filled with a volatile liquid whose vapor pressure increases as the liquid is heated; the remainder of the space is filled with the vapor. *Vapor pressure* is a measure of the tendency of the liquid to evaporate. In each case, the filled-system pressure increases with increasing process temperature. The thermal characteristic of the fill fluid is known, so the indicator has temperature graduations marked accordingly.

• *Other.* In addition to the above types of temper-re sensors, there are the following:

(1) Pyrometer. This is for very high temperatures like those of molten metal or the inside of a furnace.
(2) Color-change paint.
(3) Fusible shapes.

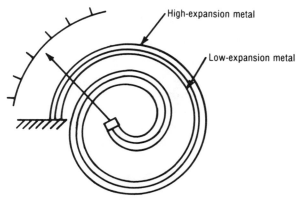

(a) Basic design of bimetallic temperature indicator

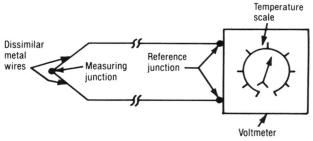

(b) Basic design of thermocouple system for measuring temperature

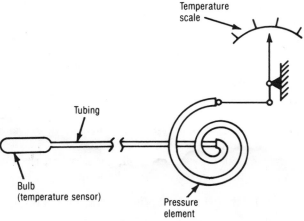

(c) Basic design of filled-system temperature sensor

The sensing system — bulb, tubing, and pressure element — is filled with fluid and hermetically sealed

Figure 3-21. Temperature Sensors

3-8. HOW ANALYSIS SENSORS SENSE

3-8-1. Units of Analysis

Analysis sensors, or simply *analyzers*, determine the composition of a *mixture* or a *solution* of different substances or the concentration of a particular substance.

• A *mixture* is an intermingling of two or more different substances that are not bound to each other and that retain their individual characteristics. The substances merely share a common space. Examples are a mixture of water and petroleum oil, water and sand, and fly ash in smoke from a fire. Mixtures can be treated to separate their components, for example, by filtering the sand out of the water.

• A *solution* is an intermingling and merging of two or more different substances so that their individual characteristics disappear and the different substances are not distinguishable; the solution is uniform throughout. For example, let us add a small quantity of table salt to a glass of pure water. The white salt dissolves and disappears from view. The liquid looks unchanged but it has become a solution, and its properties have changed. Its freezing point is lowered, its boiling point is raised, it conducts electricity better, its density is raised, and its viscosity and still other properties are changed. Continue to add salt. A point is reached where no more salt can dissolve, and we say that the solution is now *saturated*. Add a tiny bit more salt; it will not dissolve but will fall to the bottom. We now have (a) a salt-and-water solution and (b) a mixture of solid salt with the solution. We can filter out the crystals from the mixture. If we then heat the solution, the water boils off and leaves behind all the salt that had been dissolved.

Analyses are generally stated as a *concentration* in one of the following ways:

• Percent by weight or percent by volume.
• Parts per million (ppm).
• Parts per billion (ppb).

An important exception in stating analyses is the measure of acidity, which depends on the *activity* of hydrogen ions in a liquid. (Water is made up of hydrogen ions and hydroxyl ions.) Activity is related to concentration and is usually stated as pH, which stands for *the power of the hydrogen ion.*

Pure water at specified conditions has a pH of 7.0, acids have a pH of less than 7.0, while alkalies — the inverse of acids — have a pH of more than 7.0. The pH of orange juice is approximately 3; of milk, 6; of magnesia, 10.

3-8-2. Types of Analysis Sensors

Analysis measurements are inferential, that is, they depend on properties of the particular substance we are analyzing (see Section 3-1-1). A measurement of these properties tells us the concentration of the substance. For example, the concentration of salt in a water solution can be determined by measuring either the density of the solution, its ability to conduct electricity, optical characteristics, or other properties. In many cases, a particular analysis can be performed by more than one kind of analyzer.

The analyzers mentioned below are used for continuous analysis of a process and, aside from servicing requirements, operate without human attention. There are also laboratory-type analyzers and techniques that are for use by a control chemist in a laboratory.

● *Infrared.* Infrared rays have frequencies lower than those of the red end of visible light and higher than those of microwaves. With few exceptions, when infrared radiation has a frequency that oscillates over a range, some of the radiated energy is absorbed by chemical molecules and increases the motion of the molecules. For different chemicals, the molecules absorb different combinations of frequencies of the radiation. The pattern of absorbed frequencies identifies the molecules in the sample. The amount of absorption of the rays at specific frequencies is a measure of the concentration of the molecules.

Infrared analyzers are used for gases and liquids.

● *Thermal Conductivity.* Thermal conductivity is the ability of a substance to conduct heat. Different substances conduct heat at different rates. In practice, what is determined is not the actual thermal conductivity of the sample in question but rather its conductivity compared to that of air at specified conditions. At 212°F, air has a conductivity of 1.000 and the conductivity of ammonia, for example, is 1.084.

Thermal-conductivity analyzers are used for gases and vapors.

● *Acidity-Alkalinity Sensing.* As explained in Section 3-8-1, the degree of acidity or alkalinity of a liquid is stated by the pH of the liquid. pH is measured by a probe that uses a special kind of glass that is sensitive to hydrogen ions. This glass is wetted on the outside by the process liquid, and on the inside by a solution of constant pH. This probe, which has a wire immersed in the internal solution, is known as the *measuring electrode.* A second probe, known as a *reference electrode,* has an internal wire that is immersed in a special solution that maintains a stable electrical potential. The wires from the two probes are connected to a circuit that measures electrical voltage, which represents pH.

● *Chromatography.* A chromatograph passes a multicomponent sample, mixed with an inert carrier gas, through a tube. The sample may be liquid, vapor, or gas but the chromatograph converts the liquid to the vapor form. The tube is filled with absorbent materials that delay the passage of the various sample components for different time periods. A sensor detects each component as it emerges from the tube. The chromatograph can present the analysis in graphical form or digitally, and can transmit an output signal.

● *Other.* Other types of analyzers include the following:

(1) Electrical-conductivity meter.
(2) Mass spectrometer.
(3) Refractometer.
(4) Analyzers based on oxidation-reduction potential, turbidity, ultraviolet radiation, and titration.

3-8-3. Sampling Systems

An analysis system cannot be better than the process sample that it analyzes. Erroneous chemical analyses caused by improper sampling may be damaging monetarily or physically, or both. This is illustrated by the following analogy.

The validity of the results of a market survey or a political poll based on a sample of the population depends on how well the sample represents the entire buying population or voting population of interest. A bad sample leads inevitably to a bad judgment. Before the 1936 presidential election, the then-flourishing magazine, *Literary Digest,* conducted a very large nationwide poll and publicly predicted that Alf Landon would defeat Franklin D. Roosevelt for the presidency. Unfortunately for the magazine, they understated Roosevelt's vote by 19.3 percent and Roosevelt was reelected in a landslide. The Gallup and Roper polls correctly predicted the winner. The magazine had used mailing lists of telephone subscribers and automobile owners, who formed a sample that was greatly unrepresentative of the general voting population, especially over half a century ago and during the Great Depression. The magazine was so discredited that it soon went out of business.

In general, a process sampling system must perform some, but not necessarily all, of the following operations:

(1) *Obtain a representative sample.* This may require the following:

(a) Select a proper location for the sample point — in some cases more than one sample point — in order to provide a representative sample.

(b) Design a sample probe for the specific application. In most cases, the probe is merely a pipe used to withdraw a process fluid sample to be sent to an analyzer. In some cases, the probe has a sensor that performs the analysis inside a process pipe or vessel; no sample is withdrawn. In either case, the probe should be placed where it will be in contact with a representative sample.

(c) Design a tubing or piping system to transport the sample to the analyzer. The possibility that the sample composition may change in transit should be taken into account. Also, the sample time lag should not be excessive for the particular application.

Figure 3-22 illustrates different transportation methods: Detail (a) is the basic method, (b), (c), and (d) are methods for decreasing the sample lag. In considering whether to conserve or dispose of the spent sample and the bypassed fluid, the economics of conserving and the environmental effects of not conserving need to be taken into account.

(2) *Reduce the sample temperature* to a value that the sample system requires.

(3) *Reduce the sample pressure* to a value that the analyzer and auxiliary devices require.

(4) *Treat the sample* as follows:

(a) Filter out or trap debris and other undesirable solids.

(b) Adsorb or trap undesirable gases or liquids.

(c) Evaporate or condense the sample to put it into the physical form —liquid, vapor, or gas — required by the analyzer, and adjust the temperature as required by the analyzer.

(d) Adjust the sample flow rate to keep the sample lag to an acceptably small value; this may require a large flow. At the same time, the withdrawal of perhaps costly sample material and thermal energy from the process should be kept to an acceptable minimum. In some cases, the removal of an additional one-tenth of a gallon per minute of process sample beyond what is required may be surprisingly costly over a year's time.

(5) *Maintain the sample in the proper condition* for analysis, and deliver it to the analyzer.

(6) *Recover or dispose of the spent sample* and the bypass fluid.

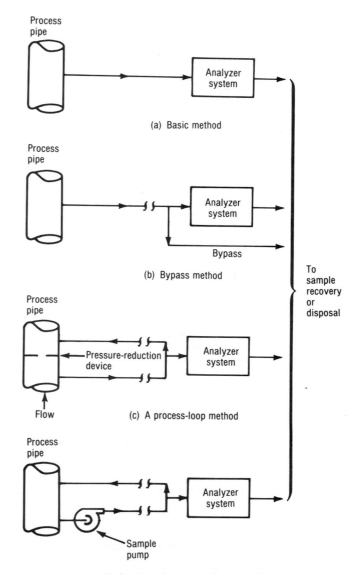

Methods (b), (c), and (d) are used to reduce excessive sample time lags.

Figure 3-22. Sample-Transportation Methods

The whole sampling system is depicted in basic form in Figure 3-23.

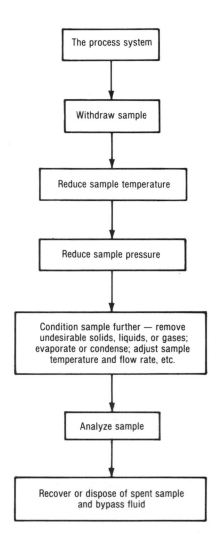

Figure 3-23. Basic Design of Sampling System

3-9. SENSORS FOR MISCELLANEOUS PROCESS VARIABLES

The previous sections described representative ways to measure the most frequently encountered process variables: pressure, flow, level, temperature, and analysis. However, process plants have many other process variables that need to be measured and controlled for proper operation. These include electrical conductivity, power, time, events, position, radioactivity, vibration, viscosity, and others.

3-10. HOW MEASUREMENTS ARE COMBINED

Sometimes we have to obtain information that we cannot measure directly but that we can calculate from other information. Suppose we need a system to measure the rate of heat input to a furnace that burns fuel gas. If we can continuously measure the heating value per cubic foot of the gas as it flows and also measure the flow rate in cubic feet per hour, then the heat input rate equals the heating value per cubic foot times the number of cubic feet per hour. The result is heat input per hour. This is accomplished as shown in Figure 3-24.

Similarly, the value of other process variables can be calculated automatically and used for automatic control, the operator's information, cost accounting, and other requirements.

ANSI/ISA Standard S5.1 includes a standard set of symbols for a variety of mathematical functions.

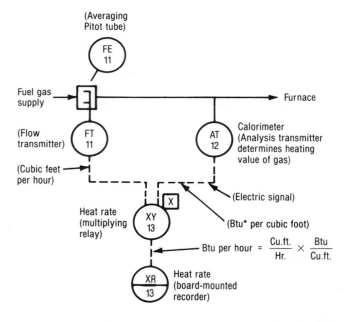

*Btu stands for British thermal unit, the amount of heat needed to raise the temperature of one pound of water by one degree Fahrenheit at specific conditions.

(The words in parentheses are for explanation and do not normally appear on a diagram.)

Figure 3-24. Measuring the Rate of Heat Input to a Furnace

4

How
Controllers
Control

This chapter covers control by controllers. Chapter 6 covers control by logic.

Every process system has a flow of material, energy, or both. The flow of material or energy is manipulated at the command of a controller whose purpose is to keep a process variable at a desired value; this value is termed the *set point*. [21] For example, a pressure controller manipulates a flow of gas, a level controller manipulates a flow of water, and a temperature controller manipulates a flow of electrical energy.

When the controlled system is at steady state, the inflow of material or energy equals the outflow. If the flow balance is disturbed, for example, if more material or energy goes out than comes in, then the controlled variable drifts away from its set point and the controller must call for a corrective action.

4-1. CLOSED-LOOP VERSUS OPEN-LOOP CONTROL

Chapter 2 describes how various instruments are connected together to form control loops. Figure 4-1 shows a generalized form of the control loops of Figures 2-1 and 2-2. All these figures illustrate automatic control information traveling around a closed circuit. This arrangement is termed *closed-loop control*.

The set point, which is adjustable, is put into a summing unit, typically by a plant operator. The measurement of actual value of the controlled process variable is also put into the summing unit. This unit subtracts the actual value from the set point, and the resulting output represents the *control error* — the amount by which the value of the controlled variable deviates from what it should be. This error is what the controller tries to eliminate or minimize in closed-loop control.

The summing unit, in most cases, is integral with the controller but is sometimes separate.

The measuring elements, which may be separate or combined, are called *feedback elements* because they feed a report on the state of the controlled variable back to the summer so that corrective action may be taken.

Most industrial control loops are of the closed-loop type. Another type of control loop is like the closed-loop except that it has no feedback, the control circuit is not closed. This arrangement is known as *open-loop control*, which is shown in generalized form in Figure 4-2.

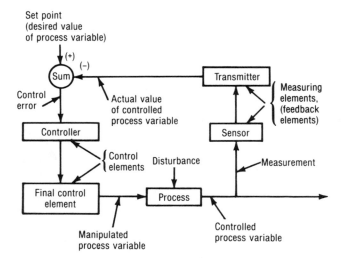

Figure 4-1. Generalized Closed-Loop Control

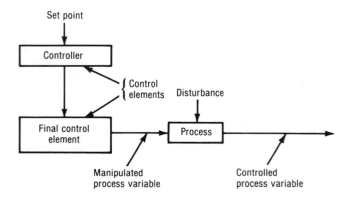

Figure 4-2. Generalized Open-Loop Control

An example of open-loop control is a person driving a car and trying to stay exactly at the highway speed limit. He steps on the gas pedal too hard, then not hard enough, and he frequently readjusts. The driver decides that he has had enough of open-loop control. He switches on the automatic, closed-loop, cruise control (see Section 2-1); ah, now he can relax (so far as the speed is concerned).

Similar cases of open-loop control exist in a process plant. Under manual control we may fill a tank, maintain a temperature, match a fluctuating flow, etc. We have to keep alert. Closed-loop control automatically does these things more reliably, usually better, and it is easier on us.

When open-loop control is used, it is often for manual operation, but it can also be performed automatically. However, it still has no feedback or else it would not be open-loop. An example of this is a chemical feed system for which we wish to use 5 gallons of reagent *B* for every 12 gallons of reagent *A*. We set up an automatic proportioning system to do this but the system does not check on itself to make sure that the proportion of the reagents is correct.

We may have an alarm to warn us when the mixing is not correct; that would provide information feedback but not automatic control feedback. In such a case, the loop can be said to be closed through the plant operator, who will perform as a feedback control element. But this is not what is meant by "closed-loop control".

Typically, an automatic controller has a built-in manual switch to cut out the automatic control and permit manual control of the output signal.

4-2. HOW AUTOMATIC CONTROLLERS FUNCTION

For an automatic closed-loop controller to keep a process variable at its set point, it must know whether the variable is at the correct value. But a YES or NO answer is not enough; the controller must also know, as a minimum, whether the value of the variable is too high or too low. For better control, it must know how much too high or too low the value is, in other words, the magnitude of the control error. For even better control, it may need to know for how long the error existed. For still better control, it may need to know how fast the variable is changing.

These various refinements of control make up different *control modes*, which are described below, beginning with *binary control*. A control mode is a specific way in which the input of a controller affects the output. [80,81]

4-2-1. Binary Control

Binary control — also known as *on-off control* and *two-position control* — is the simplest of all the control modes. It has shortcomings and is the least adaptable of all the types of control, yet it is completely adequate in many cases and is frequently used in process plants. It is also the least costly control and is the type almost always used in home thermostats.

The output of a binary controller is either ON or OFF. Its value depends on the following factors:

- The *direction* of the control error.
- The controller action, direct or reverse.

The summing unit of Figure 4-1 determines the direction of the control error, positive or negative.

Section 2-2 describes how a binary controller operates in an electric iron. For an industrial process, consider a tank of oil whose temperature is controlled by a binary controller. The oil is heated by steam in a pipe located inside the tank. If the temperature is low, the controller opens a steam valve fully; if the temperature is high, the valve closes fully. There is no intermediate flow of steam — the flow is either zero or 100 percent.

In binary control, the process difference, the operating band, between the ON and OFF actions is usually small, as, for example, in room temperature control. However, there are cases where the on-off band is intentionally made large in order to minimize the frequency of operation of equipment, such as a pump. This is frequently done for sump level control. A *sump* is a pit into which liquids drain by gravity. When the sump becomes full, a binary level controller starts a pump that operates until the sump is empty. The controller then stops the pump, and the sequence repeats. This kind of binary control is known as *differential gap control*.

Binary control is simply a switching action, and the same instrument can be used for any other switching operation, for example, to operate an alarm. Then the instrument is tagged as a *switch*, not as a *controller*.

Almost all control systems operate with varying process loads. The *process load* is the amount of material or energy that must be manipulated in order to control the controlled variable. The load depends on whether the system is operating at capacity or less than capacity, on the weather, on the quality of the plant feed, and other factors.

Binary control is theoretically and actually incapable of keeping the controlled variable within a given operating band for more than one process load. At other loads, the band moves up or down, depending on whether the load has increased or decreased. The shift of the control band is called *offset* or *droop*, which is small or large, depending on the size of the load change.

The offset is a reason why a thermostatically controlled house that is poorly insulated requires the set point to be raised manually in order to keep the house comfortable when the weather turns colder. This thermostat is a binary controller whose operating band has fallen because the house heating load has increased.

4-2-2. Proportional Control

Proportional control, also known as *single-mode control*, provides a modulated output that can have any value from bottom to top of the output range. The value depends on several factors: the direction *and magnitude* of the control error, the *gain*, or sensitivity, of the controller, and the controller action, either direct or reverse.

Controllers and processes, like people, are sensitive to a greater or lesser degree. If we say the wrong thing to some people, they flare up in anger; we consider them very sensitive. Say the same thing to other people and they may laugh it off or ignore it; they are less sensitive. The greater the sensitivity, the greater is the reaction to an input stimulus.

Sensitivity is normally expressed in terms of *proportional gain* or often simply *gain*, which is equivalent to the amount of reaction divided by the amount of stimulus. *Gain* is defined as the change of output corresponding to a given change of input divided by the change of input. [22] For most controllers, the proportional gain is adjustable. (Instead of the preferred term, *proportional gain*, the opposite term, *proportional band*, is sometimes used. *Proportional band* is the input change divided by the output change and is also usually stated as a percentage.)

A controller whose output changes by 10 percent of its initial value when the input changes by eight percent of its initial value has a gain of 1.25 or 125 percent. If the same input change causes an output change of four percent of its initial value, the controller gain is 0.50 or 50 percent.

The word *gain* applies also to instruments other than controllers and to complete loops. The input to a control valve typically comes from a controller. If the valve input signal changes from 5 psig to 7 psig (an increase of 40 percent) and causes the output (which is process flow) to change 20 percent from 100 gpm to 120 gpm, the valve gain is 0.50 or 50 percent or 10 gpm per psi.

Similarly for processes: If the steam input to a water heater changes from 10,000 pounds per hour to 12,000 pounds per hour, causing the water temperature to change from 180°F to 200°F, the heater gain is 0.010 or 1.0 percent or 1°F per 100 pounds per hour.

Now replace the binary controller of Section 4-2-1 by a proportional controller. If the oil temperature is a little too low, the controller makes the steam valve open a little; if too low by a large amount, the valve opens farther. If the temperature is a little too high, the controller makes the valve close a little; if too high by a large amount, the valve closes farther. Under normal conditions, the controller output is modulated, the control valve modulates the steam flow, and there is generally some flow.

Like a binary controller, a proportional controller is subject to offset and is incapable of keeping the controlled variable at its set point at more than one process load. The offset is the reason why a thermostatically controlled automobile cooling system runs cooler in cold weather than in hot weather. In this case, the thermostat is a self-actuated regulator that provides only proportional control.

4-2-3. Proportional-plus-Integral Control

Proportional-plus-integral control is also known as *two-mode control*, *PI control*, and *automatic-reset control*. The integral mode is also called *floating control.* Just as for proportional control, the controller output is modulated, but the value of the output depends on the following factors:

- The direction, magnitude, *and duration* of the control error.
- The gain of the controller, which depends on the proportional gain *and the integral gain*, both of which are adjustable.
- The controller action, direct or reverse.

Now let us use a PI controller for the oil temperature. This controller performs just as the proportional controller does but it has an added feature. Assume the oil temperature is low. The proportional mode output calls for an increase in steam flow, which is supposed to bring the temperature up to the set point. If the temperature remains low, the integral mode gradually adds to the output and the valve opens further. So long as a control error remains, the integral mode keeps adding to the output, the valve keeps opening, and the temperature finally reaches the set point. The control error is now zero, so the controller is satisfied; its output remains constant and the steam flow remains constant until the oil temperature is disturbed again. Then the control system again responds to the deviation from set point. The additive action of the integral mode reinforces the action of the proportional mode in either direction, up or down.

Because of the integral mode, this controller does not have control offset. At any new steady load, the controller returns the controlled variable to its set point. (Compare with Sections 4-2-1 and 4-2-2.)

4-2-4. Proportional-plus-Integral-plus-Derivative Control

Proportional-plus-integral-plus-derivative control is also called *three-mode control* and *PID control.* The derivative mode is also called *rate action*. A three-mode controller modulates its output, whose value depends on the following factors:

- The direction, magnitude, duration, *and rate of change* of the control error.
- The gain of the controller, which depends on the proportional gain, the integral gain, *and the derivative gain*, all of which are adjustable.
- The controller action, direct or reverse.

A PID controller used for the above example of oil temperature control adds the following feature to those of the PI controller: The derivative mode changes the controller output according to how fast the control error is changing. If the error is not changing, the derivative mode does nothing. If the error changes slowly, the derivative mode changes the output a little during the error change. If the error changes rapidly, the derivative mode makes a large change to the output during the error change. The purpose of the derivative mode is to give a boost to the corrective action of the controller, especially when the controlled variable is changing fast. Thereby the derivative action tends to prevent the error from becoming excessive before the proportional and integral modes can regain the upper hand. The derivative mode is used mostly for slow systems.

To control our oil temperature, let us use a three-mode controller. If the temperature fall below the set point is slow, the controller acts very much like a two-mode controller. If the fall is rapid, the action to increase the opening of the steam valve comes mainly from the proportional and derivative modes. This reduces the rate of fall and causes the derivative action to become less important. In the meantime, because of the prolonged deviation from set point, the integral action, which responds to the duration of the deviation, becomes more important. Finally, the temperature is brought back to the set point and stays there, the derivative effect has dropped to zero, and the controller output and the steam flow hold steady at the set point.

4-2-5. Process Controllability

The four control modes we have discussed — binary, proportional, integral, and derivative — are the ones commonly used. Their responses are summarized in Table 4-1.

The choices of binary control or modulating control, and which type of modulating control to use, depend on the difficulty of controlling the process. The easier a process is to control, the simpler the controller can be. In fact, a process that changes very infrequently and is otherwise easy to control may be handled satisfactorily by manual control, assuming that the possibility of misoperation or nonoperation because of human failure is taken into account. On the other hand, the more difficult the process is, the more sophisticated the controller and its adjustments generally must be.

Factors that affect the ease of control are stated in Table 4-2, and are explained below. Process factors that influence process controllability are items 1 through 10. Instrument factors are items 11 through 13.

Table 4-1. Controller Responses to Errors

Control Mode		Responds to the Control Error			
		Direction	Magnitude	Duration	Instantaneous Rate of Change
Binary (on–off)		●			
Modulating	Proportional (P)	●	●		
	Integral (I)	●	●	●	
	Derivative (D)	●	●		●
	PI	●	●	●	
	PD	●	●		●
	PID	●	●	●	●

Table 4-2. Factors Affecting Process Controllability

Increasing Process Factor	Effect on Ease of Control	
(1) Permissible deviation from set point	Helps	
(2) Load-change magnitude		Hurts
(3) Load-change rate		Hurts
(4) Process nonlinearity		Hurts
(5) Resistance		Hurts
(6) Capacitance	Helps	
(7) Dead time		Hurts
(8) Process noise		Hurts
(9) Environmental change		Hurts
(10) Control valve pressure drop	Helps	

Increasing Instrument Factor	Effect on Ease of Control
(11) Measurement nonlinearity	Hurts (in most cases)
(12) Incorrect valve flow characteristic	Hurts
(13) Signal noise	Hurts

Process Factors:

(1) *Permissible Deviation from Set Point.* Close control of the controlled variable is sometimes important, sometimes not. The looser the requirement, the greater the acceptable deviation from the set point and the easier is the control. Particularly in level control, a wide deviation of the level is often planned in order to enable the tank to absorb uncontrolled flow surges into or out of the tank without causing the manipulated flow to surge correspondingly; this type of control is called *averaging control.*

(2) *Load-Change Magnitude.* A process has a flow of material or energy. The larger the range of changes of the flow between minimum and maximum, the more difficult is the control.

(3) *Load-Change Rate.* If the flow of material or energy changes rapidly, the control is more difficult.

(4) *Process Nonlinearity.* For a discussion of this, see Section 5-1-2 covering control valve flow characteristics.

(5) *Resistance.* This is the characteristic of impeding the flow of material or energy. All materials permit heat to pass through them, but not equally well. Some materials, such as building bricks or a wool coat, conduct heat poorly; they have high resistance and are called *thermal insulators.* Other materials, such as a metal spoon or a copper wire, conduct heat well; they have low resistance and are good *thermal conductors.* The thermal resistance depends on the type and the amount of material in the path of the energy flow.

Similarly for electricity, there are insulators and conductors.

The flow of materials, also, may be inhibited by a resistance. Just as a narrow street impedes the flow of traffic, so is a small-diameter pipe more of a hindrance to liquid flow than a large-diameter pipe is. (The flow resistance of pipe is related to item 10, below.)

The greater the resistance, the more difficult it is to control a process because the resistance tends to inhibit or distort corrective control actions.

(6) *Capacitance.* This is a property of storing material or energy. It is defined as the change in the amount of material or energy needed to make a unit change in a process variable. For example, capacitance is the number of gallons of water required to change a tank level by one foot. In other words, to make a change in a controlled variable, some quantity of a manipulated variable must be supplied or removed; that quantity divided by the change is the capacitance.

Capacity is the maximum quantity of material or energy that can be stored in a device or system.

Figure 4-3 shows two tanks having the same capacity but different capacitances. So far as level control is concerned, the process is more sensitive in the tall tank than it is in the short tank because the capacitance of the tall tank is smaller; for the same change of volume in both tanks, the level in the tall tank will have a larger change. It has a faster response but is more difficult to control within a given range. Increasing the capacitance for a given control range improves the stability of the controlled variable in response to an upset.

Capacities and capacitances for typical process variables are summarized in Table 4-3.

(7) *Dead Time.* This is a period of delay between two related and sequential actions, such as the beginning of a change of input and the beginning of a resulting change of output. [23] Dead time may be called *transportation lag* or *delay time.*

Suppose a person wants to take a shower. He stands

One foot of height equals 90 gal./ft.

(a) Tall tank

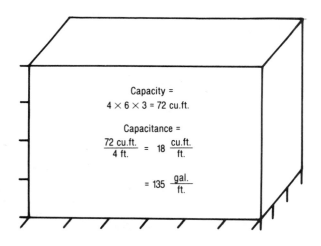

One foot of height equals 135 gal./ft.

(b) Short tank

For the same volume change of liquid in both tanks, the level in the short tank is more stable.

Figure 4-3. Tank Capacity and Capacitance

aside and opens the hot water faucet. Water comes out immediately but, unfortunately, it is very cold. Twelve seconds later, the water starts to become warm, then hot. He adjusts the temperature and steps under the shower.

Why does the person have to wait so long for warm water? It is because the shower is at the opposite corner

Table 4-3. Units for Typical Process Capacities and Capacitances

Controlled Variable	Capacity	Capacitance
Chemical Analysis	Pounds	Pounds/Percent Composition
Liquid Level	Gallons	Gallons/Foot
Pressure	Cubic feet	Cubic feet/psi
Thermal	Btu	Btu/Degree

of the house from where the hot water tank is. It takes 12 seconds for the hot water to travel from the tank to the shower. These 12 seconds are the dead time.

Similar situations occur in plant processes where the results of an upset or operating change must first be sensed, and then a flow of material or energy must be changed accordingly in order to achieve a control result. Dead time in any part of the control loop degrades the control because corrective action cannot begin during the dead time period. In the meantime, the control error has been growing.

In most cases, the dead time is not large and is not really troublesome. In some cases, especially for analysis control, the dead time can create significant difficulties that require sophisticated control schemes to overcome them.

(8) *Process Noise. Noise* is a persistent disturbance that obscures or reduces the clarity or quality of a measurement. [24] If you and I are conversing where there is much audible noise, we find it difficult to understand each other's message.

Similarly, there can be a noisy flow or level that we wish to measure. The process noise is created by fluid turbulence, high velocity, eddy currents, waves, or high velocity, which cause a rough and bumpy measurement even when the flow rate is steady. (There may also be audible noise but this does not affect the measurement, except possibly for certain process instruments that sense sound.) The greater the process noise, the more difficult the control.

To improve the control of noisy processes by quieting the noise, a fixed resistance or other suppression device may be added in the sensing line or the measuring circuit.

(9) *Environmental Change.* Changes from a sunny sky to a cloudy sky, day to night, and winter to summer affect the control of some processes because of the resulting changes of cooling water temperature or of the thermal effects on outdoor process equipment. Changes in the wind or atmospheric humidity affect the performance of equipment that cools water by evaporation.

Changes of barometric pressure affect gage-pressure instruments and thus affect operating values, especially at very low process pressures.

Such changes affect the process load, with control effects as noted in items 2 and 3. The disturbance shown affecting a process in Figure 4-1 may be a change in the environment.

(10) *Control Valve Pressure Drop.* For a discussion of this with respect to controllability, see Section 5-1-2.

Instrument Factors:

(11) *Measurement Nonlinearity.* The gain, or output-to-input relationship, of a sensor may or may not be linear, as described under "Differential Pressure" in Section 3-5. Nonlinearity is caused by the fundamental relationship of a process variable to its sensor output and is apart from minor imperfections of an instrument. A signal converter can be added to a loop to change a signal from nonlinear to linear.

Figure 4-4 illustrates output/input relationships for linear and nonlinear sensors.

(12) *Incorrect Valve Characteristic.* For a discussion of this, see Section 5-1-2.

(13) *Signal Noise.* Signal noise, particularly prevalent for electrical signals, degrades the quality of the

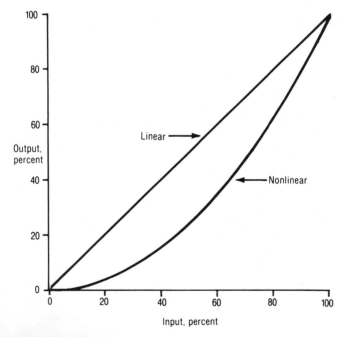

Linear: Typical for pressure or level measurement, and others

Nonlinear: Typical for differential-pressure-type flow measurement

Figure 4-4. Instrument Linearity

information carried by the signal, thereby degrading the effectiveness of the control loop. Electrical noise may be caused by interference from other electrical signals, power lines, and electrical machines. Everyday examples of signal noise are radio static from the atmosphere or visual interference on a TV screen caused by a fluorescent lamp, an electric motor, or a nearby amateur radio broadcast.

To cope with electrical noise, precautions should be taken with the instrument wiring. (These precautions include the use of shielding and grounding, twisted wire pairs, separation from potential sources of trouble, and proper terminations of the wiring.)

4-2-6. Controller Tuning

A modulating controller must be tuned for its specific service. *Tuning* is the procedure of adjusting the sensitivity of each of the controller's control modes plus auxiliary dynamic elements, if any, (see *lead-lag relay* in Section 4-3-5) so that the control system gives the best possible performance. There are mathematical procedures and process studies that can be used to estimate the best preliminary tuning adjustments for a given controller. Practically speaking, the controllers are adjusted in the field by trial and error and by experience. Even when the sophisticated methods are used, the resulting tuning should be confirmed by field trial.

Off-the-shelf electronic controllers are now available with the ability to tune themselves automatically.

4-3. COMBINATION CONTROL SYSTEMS

The basic types of control loops that are shown in Figures 4-1 and 4-2 are used in several important combinations, as follows:

- Ratio control
- Cascade control
- Selective control
- Split-range control
- Feedforward control

Each of these combinations can be used in conjunction with one or more of the other combinations.

4-3-1. Ratio Control

Ratio control is control of the ratio of two process variables. One variable fluctuates according to requirements

of the process and is called the *wild variable*. The other variable is proportioned to the wild variable and is called the *manipulated variable*. A process such as blending gasoline uses ratio control to proportion a number of components, using different ratios for summer gasoline and winter gasoline.

Figure 4-5 shows how chemical *A* is blended with chemical *B* in a desired proportion. *A* is the wild stream, *B* the manipulated stream. The two flows are measured by two transmitters that inform a ratio-computing relay what the instantaneous flow rates are. The relay divides a pneumatic signal, for example, representing flow *A* by the signal representing flow *B*, and sends the resulting ratio signal to a controller. The internal set point of the controller has been adjusted to call for the *B* flow to be 5/12 as much as the *A* flow. The controller continuously adjusts the flow of *B* so that the blend of *A*-plus-*B* contains 5 parts of *B* for every 12 parts of *A*, no matter how the *A* flow varies.

This control system manipulates stream B so that the flow of B is 5/12 the flow of stream A. The controlled variable is the blend ratio of A to B.

Figure 4-5. Flow-Ratio Control

4-3-2. Cascade Control

Cascade control is control in which the output of one controller establishes the set point for another controller. [25] Consider an oil heater that uses steam in a heating coil to heat a stream of oil, as shown in Figure 4-6(a). This system would provide good control of the oil temperature if the supply pressure of the heating steam, the oil flow, and the oil inlet temperature were all constant. Then the temperature controller would have the control valve pass exactly the correct amount of steam to keep the oil outlet temperature at the set point, and thermal equilibrium would exist in the heater.

(a) Basic scheme — single control loop

(S.P. = set point)

(b) Cascade control — two control loops

Figure 4-6. Evolution to Cascade Control

However, the steam supply pressure fluctuates widely because of a fluctuating use of steam by other users. The rate of heat transfer from the steam to the oil depends on the difference between the steam temperature and the oil temperature. The steam temperature depends on the steam pressure in the heater; the higher the pressure, the higher the temperature. When the steam supply pressure rises or falls, the immediate effect is to change the steam pressure inside the heater. This increases or decreases the steam-to-oil temperature difference, and the oil correspondingly becomes too hot or too cold. No correction is made until the temperature controller finally begins to react. In the meantime, there has been a temperature control error. Temperature control is usually slow, so the error may become large and may last for a long time.

The temperature control can be improved greatly by the system of Figure 4-6(b), When the steam supply pressure shifts up or down and causes the pressure to change inside the heater, the pressure controller senses the heater change and readjusts the control valve to bring the heater pressure back to its set point at that time. The pressure control loop acts fast, and it restores the thermal equilibrium with only a slight and brief disturbance to the oil temperature. Thus, the effect of steam pressure fluctuations is taken care of nicely for a given heater load.

But what happens if the heater load — the heat transfer requirement — changes? The load depends on the oil flow and its inlet temperature. The temperature controller feels the change in the hot oil temperature and tells the pressure controller to hold to a new steam pressure. The pressure controller is provided with a set point element that accepts a command signal. The pressure controller quickly does as it is told to provide the new steam pressure, and it continues to act promptly all by itself to correct the pressure whenever it fluctuates. However, if the load changes are severe and the control requirements are tight, additional measures may be required, as described in Section 4-3-5.

The cascade control system has two feedback controllers but only one final control element. In the example, the temperature controller typifies a *primary controller* or *master controller*. The pressure controller typifies a *secondary controller* or *slave controller.* The entire secondary control loop for pressure control can be looked at simply as an elaborate final control element for the temperature controller.

A general requirement for all cascade control systems is that the secondary control loop must be appreciably faster than the primary control loop. If the secondary is not faster, it tends to destabilize the primary control loop instead of stabilizing it; a continuing, undesirable oscillation, known as *hunting*, may occur. [31]

4-3-3. Selective Control

Selective control uses two controllers and only one control valve or other final control element, just as cascade control does, but with a big difference. Cascade control has one controller that controls the other. In selective control, each controller is independent of the other. Either controller may be in control at any given time, but when one controller is in control, the other is overridden and is ineffective. The controller that is calling for the greater corrective action in a predefined safe direction is then in charge [61]; the other controller is twiddling its thumbs, figuratively speaking. The act of selecting between the two controllers is also known as *auctioneering.* [26]

The concept of selective control is clarified by the example in Figure 4-7, which shows a tank whose level is controlled by modulating a control valve in the drain line. The flow rate of the tank drain is controlled by using the same valve. There are two limiting requirements: The level should not get too low, and the flow should not get too high. When the level is low, the level controller is normally in control and cuts back the outflow. When the flow is high, the flow controller is normally in control and cuts back the outflow. At all times, the valve takes the less-open position of the commands from the two controllers.

Liquid

Tank

(Level controller)

LC
30

(Generalized control signal)

UY
32

Select
(selector relay)

FC
31

(Flow controller)

This scheme prevents excessively low level and excessively high flow.

Figure 4-7. Selective Control

The choice of which controller shall override the other is made automatically by the selector relay, which makes a smooth transition from one input signal to the other. The function of this separate relay may be built into a controller.

4-3-4. Split-Range Control

There are control loops that use one controller and two control valves, both of which modulate. Figure 4-8(a) shows a temperature control scheme for a batch process using a chemical reaction tank that requires the reaction temperature to be constant. To start the reaction, the tank must be heated; this requires a flow of steam through a heat transfer coil. Later the reaction produces heat and the tank must be cooled; this requires a flow of cooling water through another coil.

Smooth control of the temperature is provided by the following basic system: The output of the temperature controller changes gradually as the tank temperature increases. When the controller calls for the heating valve to be wide open, the cooling valve is fully closed. When the cooling valve is wide open, the heating valve is fully closed. Halfway between, both valves are closed so that there is neither heating nor cooling. Each valve fully strokes in reverse to and in sequence with the operation of the other valve. This valve arrangement is known as *split-ranging* and is illustrated in Figure 4-8(b). [27]

4-3-5. Feedforward Control

Let us look back at the oil-heating process of Figure 4-6(a), showing a simple feedback loop for controlling the hot-oil temperature. Assume that there will be large variations in (1) the flow of oil to the heater, and (2) the temperature of the entering cold oil. Each of these changes will cause sudden and large changes in the heating load, and the flow of steam must change correspondingly. The temperature controller will eventually bring the system to a new equilibrium state. But temperature readjustment is relatively slow and, in the meantime, the exiting oil will be too hot or too cold. The feedback control system is not adequate.

There is a way, *feedforward control*, that can give excellent control during these upsets. Feedforward control makes corrections to minimize, or theoretically to eliminate, the error of the controlled variable strictly on the basis of information concerning factors that would upset this variable if the corrections were not made. [28] Upsetting factors to be taken into account are changes to properties of process feed flows and other conditions that may affect plant operation.

(a) Process control system

(b) Control valve operation

Figure 4-8. Split-Range Control

For example: We are driving a car at 35 miles per hour on a level road. We are approaching a hill with a steep upgrade, but we wish to keep our speed constant. We step on the gas pedal the moment we reach the incline in order to increase the power output of the engine. In effect, we evaluate the speed upset that would occur if we were to make no operating change when we begin climbing the hill, and we increase the fuel flow at the right time by the right amount to avoid an upset to the speed. This example uses crude manual control, yet it illustrates the principle involved.

To apply feedforward control to our oil system, the control loop must calculate the steam flow — the amount of heat — required to match the cold-oil requirement at every moment and then must provide exactly that much steam. To make this calculation, a mathematical model — an *algorithm* — must be developed to simulate the

performance of the oil-heating system. The model will need the following information:

- Oil flow rate (by measurement)
- Cold-oil temperature (by measurement)
- Specific heat of the oil (a thermal property, available from reference books)
- Heat of vaporization of water at the pressure inside the steam coil (a thermal property, available from reference books) (*Heat of vaporization* is the heat required to boil water at a given pressure.)

All this information is put into a small calculating system, UY-25, which may be made up of standard commercial instruments. The control system is shown in Figure 4-9(a). UY-25 is informed promptly when the oil flow rate or the cold-oil temperature changes. The calculator does its arithmetic, determines what the steam flow should be for the new conditions, and control valve UV-25 changes the steam flow accordingly.

Though feedforward control can improve performance greatly, it has the following weaknesses:

- We cannot make a perfect mathematical model of the process and its equipment.
- Every instrument, including control valves, performs imperfectly and contributes to the functional inaccuracy of the control system.
- The design performance of the instruments, and also of the process equipment, is further degraded by environmental changes, aging, corrosion, and other factors (see Chapter 8).

The feedforward errors become intolerable if they are allowed to accumulate. However, we know that feedback control is capable of eliminating errors after they occur. Figure 4-9(b) shows temperature controller TC-29 added to provide feedback control to supplement the feedforward control. When there is a load change, feedforward will greatly reduce the initial temperature error, and the remaining error will be eliminated by feedback. This will reduce both the magnitude of the control deviations and their duration. Such use of feedforward and feedback is typical to ensure that the controlled variable has a minumum upset and ends up at its correct value.

The control system now includes a steam flow controller, FC-28, which provides flow feedback solely to increase the accuracy of the flow control. The scheme of Figure 4-9(a), lacking the flow controller, has open-loop control of the steam flow, which may be quite inaccurate because of control valve inaccuracies; the inaccuracy is often significant for the slow temperature system.

The system also includes a dynamic element, FY-26 — a *lead-lag relay* — whose purpose is to compensate for the difference between the speed of a change in the oil flow and the speed of the resulting change in heat flow from the steam. FY-26 works by speeding up or slowing down signal changes to balance the process responses; its effect is adjustable and is part of the tuning operation for the system.

Feedforward systems are used mainly for difficult-to-control processes, which are usually slow processes with much dead time. Feedforward has worked wonders in improving some processes, such as distillation, electric power generation, and chemical manufacture.

In theory, feedforward control is capable of absolutely perfect control — no control error. In contrast, feedback control is theoretically *incapable* of perfect control because its corrective action begins only after a control error already exists. Paradoxically, the theoretically inferior control, feedback, is predominant in actual practice because it is adequate in most cases and is less complex and less expensive.

Legend:
FT Flow transmitter
FY Lead-lag relay to compensate for oil-flow/heat-flow response difference
TT Temperature transmitter
UV Multivariable control valve
UY Multivariable relay to calculate steam-flow requirement

(a) Feedforward alone

Legend:
FC Flow controller
FT Flow transmitter
FV Flow control valve
FY Lead-lag relay to compensate for oil-flow/heat-flow response difference
S.P. Set point signal
TC Temperature controller
TT Temperature transmitter
UY Multivariable relay to calculate steam-flow requirement

(b) Feedforward plus feedback

Figure 4-9. Feedforward Temperature Control

5

How Final Control Elements Function

Every control loop of any kind has to manipulate a flow of material or energy (see Section 1-2 and Figures 4-1 and 4-2). Though there are different ways to manipulate the flow, the predominant method by far is by means of a control valve.

5-1. CONTROL VALVES

We are all familiar with valves from our daily use of kitchen and bathroom faucets. These are simple hand-actuated devices for turning the flow on or off, limiting the flow, or switching the flow between the bathtub faucet and the shower head.

A *control valve* is a device, other than a common hand-actuated on-off valve, that adjusts one or more passages to change the flow rate or path, or both, for one or more fluid streams. [29, 30] In process control, there are *on-off valves*, *throttling valves*, and *modulating valves*, which are described as follows.

- An *on-off valve*, which operates in binary fashion, fully opens or closes one or more flow paths. It is an *isolation valve* or *shutoff valve* that may be used actively for manipulation (see Section 4-2-1).

- A *throttling valve* is used to limit flow by adjusting the opening. The limiting of flow is called *throttling* because the flow is choked. There are two different kinds of service, non-modulating and modulating, which use throttle valves as follows:

(1) *Non-Modulating Throttle Valve.* The valve is adjustable to different positions that remain fixed until the next adjustment; an example is a kitchen faucet. The adjustment is performed by hand, either directly or by remote control, or by a binary controller. When the valve is operated directly by hand, it is called a *hand control valve (HCV)*; when operated remotely by hand, it is a *hand valve (HV)*.

(2) *Modulating Throttle Valve.* This is usually called simply a *modulating valve.* The valve is operated by an automatic modulating controller that causes the valve to modulate in turn.

The subject of control valves can be surprisingly sophisticated and subtle. As a crude measure of this, the cost of a single valve for use in a process plant may range from 5 dollars to half a million dollars (for a nuclear power plant). Size differences are only part of the reason for the cost ratio being 1 to 100,000; there are other factors, which relate to performance, construction, materials, and quality.

A control valve is an assembly of hardware that is composed of three basic sub-assemblies:

- A body that is a pressure housing through which fluid passes. The body has connections for pipe or tubing.
- A moveable *plug* that fits inside the body. The plug position changes to manipulate the flow. The plug may be a disk, a ball, or some other specialized design.
- An actuator that is connected to the plug. It can move the plug when the actuator receives motive power.

These basic elements and the principles of valve operation are generalized in Figure 5-1. There are numerous sub-parts, including bonnet, packing, seat, stem, and others.

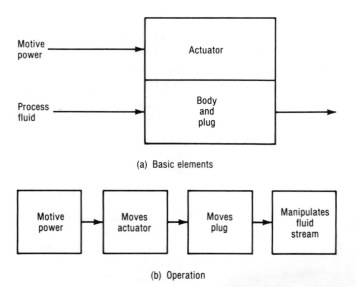

(a) Basic elements

(b) Operation

Figure 5-1. Generalized Design and Operation of Control Valves

The major considerations in choosing a control valve are discussed below.

5-1-1. Body Type

Examples of different types, *patterns*, or *styles* of valve bodies are as follows.

- *Globe.* This type has a plug that, typically, moves up and down to partially or fully open or close one or two openings, known as *ports*, as shown in Figure 5-2(a). The three-way valve in the figure is designed with two outlet connections; when the plug increases the flow to one of the outlets, it decreases the flow to the other outlet, and vice versa.

- *Angle.* This is basically like a single-port globe type except for the orientation of one of the pipe connections. It is illustrated in Figure 5-2(b).

- *Butterfly.* This has a disk or vane that is mounted on a shaft that can rotate a quarter turn, 90 degrees. The disk is like a furnace damper except that it is much more sturdy and can be built for heavy-duty service. Figure 5-2(c) depicts a butterfly design, known as a *wafer* type, having an abbreviated body for installation between a pair of pipe flanges. An alternative design has a longer body with a pipe connection on each side.

- *Ball.* Passages are cut into a ball that is retained inside the valve body. The ball is on a quarter-turn shaft. Figure 5-2(d) shows typical flow arrangements.

5-1-2. Flow Characteristic

Every valve that throttles flow has a *valve characteristic*, which is the relationship between the amount of *valve position*, *travel*, *stroke*, or *lift* in the opening direction as compared to the resulting flow. The characteristic that is best for each application depends, as will be seen, on the specific control loop and the design of the entire piping system that contains the particular valve. There are three *idealized* types of characteristic, as follows:

(1) *Equal percentage* type, for which equal increments of stroke to open the valve increase the flow by a constant percentage of the flow existing at the time of the change. The flow is not proportional to the stroke.

(2) *Linear* type, for which equal increments of stroke increase the flow by a constant amount. The flow is proportional to the stroke.

Single-seated   Double-seated   Three-way

(a) Globe

(b) Angle valve

(c) Butterfly valve

Typical flow arrangements

1st position   2nd position

Ball

Two-way design

Path

Body

Three-way design

(d) Ball valve

Figure 5-2. Control Valve Body Styles

(3) *Quick-opening* type for which equal increments of stroke give decreasing increments of flow. The flow is not proportional to the stroke.

The equal percentage and linear types require specially shaped valve internals that follow mathematical equations for the desired relationship between stroke and flow. The equations assume that the *pressure drop* —which is the pressure difference between valve inlet and outlet — is kept constant so that the valve flow depends on only the valve stroke and characteristic.

The quick-opening type generally consists of a flat disk in a globe or angle valve, as in the traditional kitchen faucet. Some manufacturers modify the ideal characteristic. The characteristic as manufactured for all the types is called the *inherent characteristic*.

Valve characteristics are illustrated in Figure 5-3, which shows that an idealized control valve that has traveled 75 percent of its total stroke has a flow of approximately 42 percent, 77 percent, or 98 percent of its rated capacity, depending on whether the characteristic is equal percentage, linear, or quick-opening, respectively.

From the figure, we see that if the valve has an equal percentage characteristic and we open the valve from 25 percent to 50 percent of its full travel, the flow increases from approximately 6 percent to 17 percent of capacity, a change of 11 percent of capacity. For a linear characteristic and the same valve movement, the flow increase is

from 25 percent to 50 percent of capacity, a change of 25 percent. Then moving the valve from 75 percent open to 100 percent causes flow changes of approximately 58 percent for equal percentage and 25 percent for linear. This is summarized in Table 5-1. Incidentally, the reason why most modulating valves do not control down to zero, as illustrated in the figure, is that there generally is unmanipulable flow due to leakage even when the valve is in the closed position.

We could go through a similar process to compare a linear characteristic with a quick-opening characteristic. However, from inspection of the figure and the table, we can draw conclusions, as shown in Table 5-2.

What is the point of having a choice of valve characteristics? After all, Figure 5-3 shows that each characteristic can handle flows from close to zero up to 100 percent of capacity. The answer is that the characteristics affect how controllers are tuned, and they can often make a significant difference in performance of the control system.

An important factor for process controllability is this: Many pumps, heat exchangers, chemical processes, and control loops operate *nonlinearly*. This means that they require varying amounts of the manipulated variable to hold the controlled variable steady as the process load changes by equal amounts. The nonlinearities exist because the process capacitances and resistances change as the load changes. The nonlinearities are different for different processes. Figure 5-4 shows how the manipulated variable varies uniformly with load for a linear process and how it may vary nonuniformly for nonlinear processes. Many control systems are more or less tolerant of nonlinearity and are not much of a problem in that respect, but cases occur where the nonlinearity is troublesome. Selecting the proper valve characteristic enables the system to be more linear, therefore easier to control.

Tuning is described in Section 4-2 as the adjusting of a controller to provide the best possible performance for the control system. If the tuned system, consisting of the process and the control loop, is nonlinear, then it is not possible to achieve the best performance for all process loads. If the controller is tuned at high load, it will be undersensitive and sluggish at low load. If it is tuned at low load, it will be oversensitive and unstable at high load. Tuning for a range of loads is easiest if the system is linear.

Another controllability factor relates to frictional pressure losses caused by flow in the piping circuit that contains the valve. Circuit pressure losses warp the valve manufacturer's inherent-characteristic curves of Figure 5-3 so that the equal percentage curve bends toward the linear curve, the linear curve bends toward the quick-opening curve, and the quick-opening curve bends more

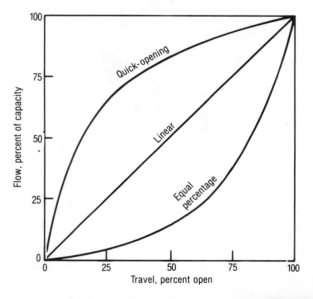

Note: When the valve is very near the closed position, its characteristic is indeterminate.

Figure 5-3. Inherent Valve Characteristics, Idealized

Table 5-1. Valve Characteristics: Equal Percentage Versus Linear

	Characteristic	
	Equal Percentage	Linear
For Valve Travel from 25% to 50% Open, the Flow Change* is	6% to 17% = 11%	25% to 50% = 25%
For Valve Travel from 75% to 100% Open, the Flow Change* is	42% to 100% = 58%	75% to 100% = 25%

*All percentages are relative to full scale.

Table 5-2. Summary of Valve Characteristics

	Valve Characteristic		
	Equal Percentage	Linear	Quick-Opening
Relative Valve Gain (Sensitivity):			
At Low Part of Range	Lowest	Intermediate	Highest
At High Part of Range	Highest	Intermediate	Lowest
Gain Change as Valve Opens	Increases	Constant	Decreases

Note:
Valve Gain equals the change in flow divided by the change in valve travel.

to the left and becomes more extreme. These altered characteristics are called *installed characteristics*.

There are rules of thumb that simplify the selection of the characteristic for the majority of applications. [34, 62] The rules are based upon the following considerations:

- The kind of control that is involved — whether for flow, pressure, temperature, level, or other.

- The ratio of (a) the pressure drop of the valve alone to (b) the dynamic pressure loss of the entire piping circuit, all at maximum flow. *Pressure drop* for a valve is the difference between its inlet and outlet pressures. The valve drop for a given piping system cannot be assumed, and it varies with load. It equals the remaining pressure difference between the start and finish of the entire flow circuit after (1) deducting the total system *dynamic pressure loss*, which is the sum of all the friction losses for the entire flow circuit, and (2) correcting for the changes of pressure due to elevation changes, if any.

The higher the ratio, the smaller is the difference between the inherent and the installed characteristics, and the easier is the control. The ratio varies with each installation.

The array of valve characteristics thus provides a choice for improving the process control by counter-acting the problems of nonlinearity and piping circuit pressure drop. However, there are remaining deficiencies in a given characteristic, as listed below, so we do not expect to end up with a linear control system that is perfectly tuned over its full operating range.

The other factors that affect the installed characteristic of a control valve include the following:

- Design approximations by the valve manufacturer.
- Manufacturing imperfections.
- Erosion of the internal parts of the valve, resulting in changes of the internal shapes.
- Later changes in design operating conditions.
- Boiling of liquid — known as *flashing* — which occurs inside the valves under some operating conditions.

There are accessory instruments, known as *characterizing relays*, that enable a nonlinear instrument signal to be linearized or otherwise shaped in order to improve system controllability or to be used for calculations. And there are control valve accessories that enable a given valve characteristic to perform as though it were different.

Each load change is from one steady state to another steady state.

For a Linear Process: If a low process load is increased (or decreased) by 20 percent of full load, for example, the process controller responds by changing the manipulated variable by some amount, just as a bicyclist must pedal harder when he comes to an upgrade. A similar 20-percent load change at high load likewise results in a change of the manipulated variable. Both manipulated-variable changes are *equal* in amount if the process is linear.

For a Nonlinear Process: Equal changes of load at different loads cause *unequal* changes of the manipulated variable.

Figure 5-4. Process Linearities and Control

5-1-3. Valve Size

For a valve, the word *size* needs clarification. *Body size* states the nominal diameter of the pipe connections, not the exact diameter. *Port size* refers to the diameter of the port. The port may be *full-size*, meaning that it has the same nominal size as that of the body. Often the valve has a *reduced port*, also called *reduced trim* or *restricted trim*, meaning that the port is smaller than the body. *Valve size* may be stated as "2 in.", for example, which usually refers to the body size. Another valve may be "2 in. × 1 in.", which means that the valve has a 2-inch body and 1-inch trim.

Mathematical equations are used to size control valves, depending on the physical state and the operating conditions of the fluid. [33] These equations express the flow capacity of a valve in terms of a valve-sizing coefficient, C_v. C_v is the number of U.S. gallons per minute of water at 60°F that can flow through the valve when it is wide open and the pressure drop is one psi. A larger C_v means a higher flow capacity. Though water is the reference fluid, C_v is used for valves handling all fluids. Manufacturers' catalogs list the C_v ratings for their various control valves according to the body style, body size, port size, and other factors. When a valve C_v is specified, it is understood that this is for a fully open valve, unless stated otherwise.

A rule of thumb for preliminary sizing states that the proper valve size is one size smaller than the pipeline size; for example, this suggests a 3-inch valve in a 4-inch line. However, the size must not be considered final until it is actually calculated from the appropriate equation. In a well-designed process system, the valve body size does not exceed the line size and it is not unusual for it to be more than one size smaller.

The 3-inch valve may be as adequate as a 4-inch valve for manipulating the flow, and it would cost less. More importantly, the smaller valve very often controls better. A valve that has a greatly excessive capacity has operating disadvantages, especially for throttling service. The oversized valve, compared to a properly sized valve, throttles with the plug nearer the closed position, causing the following bad effects:

- *Excessive Gain.* A slight stroking of the oversized valve causes a relatively large change of valve opening, with a correspondingly large change of flow. In other words, the valve gain, which is the ratio of (a) the change in flow to (b) a change in valve position, is high. High valve gain together with any erratic error in positioning the valve — and some error is expected, even with a positioner — magnify the resulting flow error; stable control is more difficult to achieve. This would be comparable to trying to have very fine control of the water flow in the kitchen sink if we replaced the faucet with a 3-inch valve.

- *Effect on Flow Characteristic.* Because of design and manufacturing limitations, uncontrollable leakage flow generally exists and causes the planned characteristic of the valve to be indeterminate at very low valve positions. This effect becomes aggravated if the valve parts erode or corrode. Leakage flow may become important when the total flow is small. In general, the larger the port size, the greater is the leakage and the worse the resulting control at low flow.

- *Dynamic Instability.* When a valve, such as a globe valve, is closed, its plug rests on that portion of the valve port known as its *seat*. In the near-closed position of the valve, the smallest area of fluid flow is the relatively small space between the plug and seat. This flow constriction causes a local increase of fluid velocity with a lowering of pressure, as shown in Figure 3-17 for an orifice plate. Especially for high pressure drops, the lowering of pressure at the valve seat can have a strong sucking effect on the plug and cause it to slam against the seat. The sucking is the so-called *bathtub-stopper effect,* which occurs when a bathtub is being drained. If a rubber stopper is placed in the tub outlet by hand to stop the flow, the stopper is forcibly sucked down when it is very close to the outlet, and the stopper slams closed. For a control valve in modulating service, such stoppage of flow eliminates the downward pull on the valve plug, and the valve actuator reopens the valve to maintain the required flow. With the resumption of flow, the plug is promptly pulled down again while the plug is still close to the seat; this close-open cycling goes on and on. This action can be mechanically damaging to the valve and is upsetting to the process. One way to overcome this effect is by installing a larger actuator to overcome the dynamic forces that cause the plug to cycle; this is a brute-force method that may increase the valve cost and may cause external space problems. It is better, where practical, to minimize the dynamic cycling by avoiding excessive oversizing of the valve.

It is sometimes advisable to provide a larger body than is calculated on the basis of capacity. This may be because of the following considerations:

(1) The fluid velocity entering or leaving the body may be excessive, thereby causing excessive erosion and noise. Enlarging the body reduces the inlet and outlet velocities.

(2) If flashing and cavitation (boiling and recondensation) occur as a result of internal velocity changes, a larger body reduces the severe erosive effect of collapsing bubbles on the body wall. Also, the wall is thicker, with a longer life.

(3) A structurally stronger body may be needed to withstand piping forces and vibration.

(4) The costs of pipe reducers and their installation can be avoided by using a line-size body.

(5) A larger body may be desired to allow for future plant expansion and a higher flow but the port can nevertheless be sized to meet current needs by using reduced trim. Our calculated 3-inch valve may become a 4-in. × 3-in. valve having a body enlarged to 4 inches but a proper 3-inch port. Whenever required, 4-inch trim can be installed in the 4-inch body, and the piping will not have been disturbed.

(6) The port may be oversized for current needs because of either an initial sizing error or a change in the valve operating conditions. The oversizing can be corrected by installing reduced trim.

Rangeability is an important concept for control valves. *Inherent valve rangeability* is usually defined as the ratio of the maximum flow that is controllable by a valve to its minimum controllable flow. The maximum controllable flow is the valve capacity, which is, of course, greater than the maximum flow requirement or else the valve will be a process bottleneck. The minimum controllable flow is not so clear-cut: It may depend on how large the leakage flow is — this gets worse with wear — or on how low the flow is where the valve characteristic becomes deviant. The inherent rangeability of types of valves can be stated only as approximations, with some common values as stated in Table 5-3. [63] A valve in actual service has an *installed rangeability* that is smaller than the inherent rangeability because of pressure losses in the piping.

Table 5-3. Nominal Inherent Rangeability of Typical Control Valves [63]

	Stroke Resolution*		
Characteristic	1%	10%	20%
Quick Opening	50:1	6:1	3:1
Linear	100:1	10:1	5:1
Equal Percentage	50:1	50:1	40:1

Stroke resolution refers to the positioning accuracy for the valve. It is of particular importance for defining the minimum controllable flow, which exists when the plug is close to the seat.

Process rangeability is the ratio of the maximum required C_v to the minimum required C_v, usually corresponding to the maximum and minimum flows, respectively, that are to be manipulated, with each flow being at its own operating conditions. When considering a control valve for a specific service, its installed rangeability should be larger than the process rangeability. Otherwise, the valve may be called on to manipulate flow over a range that is larger than it can handle. In such a case, a different style of valve may be called for, if otherwise suitable, or it may be necessary to use two control valves in parallel.

5-1-4. Actuator

The *actuator* is the drive or power element that positions the part of a final control element that directly manipulates the fluid flow or other manipulated variable. In the case of a control valve, the actuator positions the plug. (The word *actuator*, referring to the drive, is much preferred to *operator*, which should generally be reserved for a person. A person *operates* and an actuator *actuates*. However, it is also correct to say that a controller, for example, operates the actuator of a valve.) The parts of a typical actuation system are shown in Figure 5-5.

There are different kinds of actuators, based on the following requirements:

• *Type of Movement.* A globe valve, for example, requires an actuator whose output movement is *rectilinear*, or simply *linear*, to move the valve plug up and down. By contrast, a butterfly valve requires an actuator having a *rotary* output to rotate the valve disc. The valve movement begins with the drive, whose input movement is not necessarily like its output movement. The globe valve may be moved up and down either by manually rotating a knob that rotates a threaded valve stem, as is common in the home and industrially, or by pushing or pulling the valve stem, as is common industrially for control valves. The butterfly valve may be rotated by a rotating motor, usually electric, or by a push-and-pull actuator whose linear movement is converted into rotary movement by a mechanical linkage, as for a fireplace damper.

• *Control Action.* Actuators are used for on-off or modulating service. Electric actuators, such as motors and solenoids (which are basically electromagnets), may overheat and burn out depending on how frequently they are energized and how heavily and for how long they are loaded.

• *Power Output.* To move a valve plug or other device requires the actuator to provide sufficient output force, which may be considerable for the mechanical parts of large valves. Additional actuator force is needed to overcome the forces of the process fluid. The fluid forces may be very large because of (1) a static pressure difference between inlet and outlet, resulting in an unbalanced force, and (2) dynamic effects existing when there is flow. It may require more force to modulate a flow and keep the flow steady than to merely open the valve fully or close it. Another factor is how fast the valve is required to open or close. There are many processes for which speed of operation is unimportant; this may be true of a slow-changing process. In some cases, especially for emergency services involving safety, speed of operation in opening or closing may be very important. Of course, the greater the speed requirement, the more powerful must the actuator be, and the more expensive the actuator. Some turbine shutoff valves, for example, are 20-inch size or larger; handle high-pressure, high-temperature steam; and are massively large and heavy. They full-stroke closed in 0.1 second; that takes a lot of power.

The force that an actuator can apply is described as follows:

(1) For linear motion: The force may be stated as a "stem thrust of 1200 pounds," for example.

(2) For rotary motion: This needs a twisting force, which may be stated as a "torque of 600 pound-feet," for example. *Torque* is the twisting force that makes a bolt turn when we use a wrench on it as a lever. The value of 600 pound-feet for the twisting force is the product of a linear force, in pounds, multiplied by the length of the lever, in feet. The SI unit for torque is the *newton meter*, which equals 0.7375 pound-feet.

Rotary actuators may also have a power rating, such as "one horsepower," which depends on both torque and speed. When an actuator moves a valve through its stroke, the actuator force is exerted over the stroking distance. The force multiplied by the distance is the *work* done by the actuator. The work divided by the stroking time is the *power* capability or requirement for the actuator. The basic SI unit for power is the *watt*; one kilowatt equals 1.340 horsepower.

Similarly, a car that is driven up a hill at high speed requires more power than at low speed even though the work performed is the same in both cases. The work is performed by the car raising itself from the bottom of

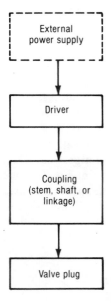

Figure 5-5. A Typical Control Valve Actuation System

the hill to the top. Work does not depend on time; power does.

Altogether, the actuator must have the power and the force to provide the following positioning results:

(1) Accuracy despite unbalanced fluid forces, the weight of parts, and frictional resistance.
(2) Stability despite process pressure fluctuations.
(3) Speed of operation as required for the application.

Following are basic descriptions of common types of actuators.

● *Fluid-Powered Actuators.* The great majority of control valve actuators in the process industries are fluid-powered using compressed air. The most commonly used air pressures are approximately 50 psig, 80 psig, or 100 psig for on-off service, with typical ranges of 0 to 20 psig or 0 to 35 psig for modulating service. In some cases, the power medium is hydraulic liquid at pressures of 3000 psig or higher. The fluid-powered category includes the types of actuators described below. Miniature versions are used for pilot valves that are sometimes used to control the operation of process control valves.

(1) *Diaphragm Actuator.* This has a flexible diaphragm with a partial backing plate that is rigid. The air pushes one way, and an opposing spring pushes the other way. A valve stem is connected to the plate. The stem position varies with the air pressure. This is illustrated in Figures 5-6(a) and (b), which show control valves that move down and move up, respectively, as the air pressure increases. These are known as *fail-open* and *fail-closed* valves, respectively, because, if control air to the actuator is lost, the (a) valve opens and the (b) valve closes.

Both of the valves are *fail-safe*, meaning that they fail in a safe direction. For example, suppose a temperature controller is modulating the fuel gas flow to a furnace and the control air line to the control valve is damaged and breaks, causing the actuator air chamber pressure to fall to atmospheric pressure. The spring makes the valve fail to the safe position.

But which is the safe position, open or closed? If the valve were to fail open, there would be an unrestricted flow of fuel gas to the furnace; this could be very dangerous. If the valve were to fail closed, the furnace would be starved of fuel and the fire would go out; production would stop but there would be no danger. The choice is clear: We specify that the valve shall fail closed, as does the valve in Figure 5-6(b). Nobody is hurt, we repair the broken air line, and we are back in business.

There are four possible modes of actuator failure: *fail open, fail closed, fail locked* (in the last position), and *fail*

(a) Fail-open   (b) Fail-closed

Valves with diaphragm actuator

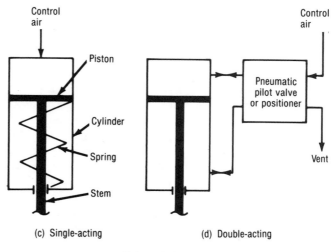

(c) Single-acting   (d) Double-acting

Piston actuators

The examples above illustrate actuators in either modulating or on-off service. On-off operation is shown also in Fig. 5-8.

Figure 5-6. Diaphragm Actuators and Piston Actuators

indeterminate (difficult to predict). ISA has established standard symbols covering the four modes. [35]

(2) *Piston Actuator.* This uses a piston and cylinder. Figure 5-6(c) illustrates one version, called *single-acting*, that uses air pressure on one side of the piston, and an opposing spring on the other side, similar to the arrangement for the diaphragm actuator, described above.

Figure 5-6(d) shows another common version, called *double-acting*, that uses no spring but has control air on both sides. This requires the use of a pilot valve or a positioner, which are accessories described below. An input signal to the accessory causes the opposing output air pressures to vary in reverse to one another so as to balance the fluid forces on the control valve plug and push the piston up or down as required.

Regarding fail-safe action, the single-acting piston actuator functions like the diaphragm actuator. But the double-acting piston requires a small auxiliary subsystem that includes an air reservoir for emergency use. There are other variations.

(3) *Motor.* This is a pneumatic rotary machine with gearing that can be used for on-off or modulating actuation of valves and other devices.

● *Electric-Powered Actuators.* Electric-powered control valves are used for process work less often than are pneumatic-powered valves. Electric-powered types are as follows:

(1) *Motor Actuator.* This is essentially an electric rotary motor with gearing that is used mainly for on-off control valves and other devices. Some are used for modulating service, especially in heating and air-conditioning systems.

(2) *Solenoid Actuator.* A solenoid is an electro-magnet that has an annular (ring-shaped) electric coil, inside of which is a magnetic iron plunger that is connected to a valve stem. Applying voltage to the coil pulls the plunger into the coil. A spring returns the plunger to its initial position when the electric power is shut off. The stem movement is linear. Thus the valve opens or closes, depending on the design, as the actuator is energized.

Though solenoid valves can be designed to modulate by varying the power supply voltage, their major use is for on-off service. Solenoid actuators produce relatively little thrust and have limited use as control valves. They are used mainly as pilot valves for process control valves.

● *Manually Powered Actuators.* A manual actuator may be a handwheel having a rim, spokes, and hub. The wheel may be modified or replaced, as follows:

(1) There may be no rim and the number of spokes may be reduced to two or even one.
(2) There may be no rim and no spokes. The hub has wrench flats for use with a wrench.
(3) The rim may have a chain to permit manual operation from a lower elevation.
(4) The actuator may include a set of gears to (a) reduce the force that the operator must exert, or (b) change the type of output motion from rotary to linear.

Manual actuators alone are used only when remote or automatic operation of a valve or other device is not required. Their major use is for isolation of equipment or systems. They may also be used to throttle flow where the process is slow-changing and their relative unreliability, because of their dependence on a human operator, is acceptable. (See "Process Reliability and Safety" in Section 1-3.)

A manual actuator may be attached to an automatic valve to override the automatic actuator. This may be done to permit continued plant operation if the automatic control system is out of order. This manual actuator has a clutch for disengaging the automatic actuator, after which the operator has full manual control of the valve.

A manual actuator, known as a *jack* or *hand jack*, may be used not to override the automatic actuator but rather to limit the range of travel of the valve. It is an adjustable travel stop. The jack is threaded and may have a hand-wheel or wrench flats.

A valve stem may be lengthened by placing a stem extension between the manual actuator and the valve body. This permits manual operation of the valve from a distance.

● *Positioner.* A major accessory for many modulating control valves is a *positioner*. This is a position controller that senses valve position directly by means of an external mechanical connection to the valve stem. The set point of the positioner is valve position, as commanded by a process controller, such as a temperature controller. The positioner automatically adjusts the signal to the valve actuator if the actual position does not match the set point. The position controller is part of a cascade control loop (see Section 4-3-2), which is illustrated in Figure 5-7.

The figure shows a system for cooling hot water by mixing it with cold water. The temperature of the tank effluent is controlled by a modulating temperature controller that controls a positioner that operates a control valve that manipulates the flow of cold water. The temperature controller is the primary controller of the cascade loop; the positioner is the secondary controller.

The valve could operate without the shown positioner but, in this particular case, it would not operate as well. The tank has a large thermal capacitance; therefore, temperature changes are slow. Because of friction and unbalanced fluid forces in the valve, a control valve without the positioner would not take *exactly* the position that corresponds to the temperature controller output signal; there is a valve-position error that causes an improper flow of cold water. The position and flow will not be corrected for a relatively long time because:

(1) The process is slow.

(2) The flow error is small.

(3) No adequate correction will be made until the temperature error has built up again to change the temperature controller output sufficiently to overcome the forces that restrain the valve movement.

Finally, the valve position is readjusted, again perhaps not exactly correctly. The point is that the control error had to keep growing for a while because the valve responded imperfectly to the command of the temperature controller. The control error will be greatly reduced by having a positioner on the valve.

The rule is that a positioner is not used unless it is needed. It is needed only for modulating control in the following cases:

(1) If the process is slow, as is common for control of temperature or analysis, for example. Control is actually degraded if a positioner is used for a fast process, such as most pressure- or flow-control systems; this is *regardless of how bad the valve friction and unbalanced forces*

may be. This follows the general rule for cascade systems, namely, that the secondary controller should be installed only if the primary loop is slower than the secondary loop (see Section 4-3-2). In this case, the positioner should be used because the control valve can change faster than the process can. This practice differs from the traditional but obsolete criterion for using a positioner to overcome forces that impede accurate positioning.

(2) If a double-acting actuator is used, regardless of the process speed.

(3) Other special situations.

Valve travel may be speeded up for a modulating pneumatic valve that does not require a positioner by adding a booster relay at the valve. The signal sequence is from controller to booster relay to valve actuator. A *booster relay,* or *volume booster,* is an instrument whose output pressure equals its input pressure but whose output air-flow capacity is larger than that of the input. Thus it speeds up the pressurizing or venting of the sizeable air chamber of a valve actuator.

Double-acting piston actuators in on-off service do not use a positioner but rather a small four-way valve, as shown in Figure 5-8(a). The valve moves the actuator fully up or fully down by switching air pressure from one side of the piston to the other. This valve may be hand actuated but frequently is solenoid actuated or pneumatic actuated. The valve is named a *pilot valve* because it does no work other than to direct the action of the control valve actuator. Figure 5-8(b) shows a three-way pilot valve used for a single-acting piston actuator.

5-1-5. Materials of Construction

There is a large choice of available materials of construction for valve bodies and their internals. The internals are known as *trim.* When specifying valve materials, the following factors must be considered: structural strength, corrosion, and erosion.

● *Structural Strength.* The material of the body and piping connections is normally specified by the purchaser. The body and connections must be made strong enough to withstand the following forces:

(1) *Internal pressures in combination with high or low temperatures.* The greater the pressure, obviously the stronger must be the material. However, the strength of any metal (a) decreases as its temperature rises, and (b) is limited by brittleness and potential fracturing at cryogenic (very low) temperatures. The American Society for Testing Materials (ASTM) has established standard

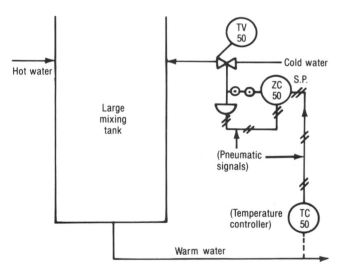

ZC (position controller) = valve positioner

—⊙-⊙— = Mechanical link, used here to sense valve position

S.P. = Set point

TV = Temperature valve

Piping-and-instrument drawings such as this do not ordinarily show the positioner. Instead, the controller output is shown going directly to the valve actuator.

Figure 5-7. Using a Valve Positioner

specifications for metals, and the American Society of Mechanical Engineers (ASME) has established allowable pressure and temperature working limits for different uses and types of metals. For example, a specific grade of carbon steel (ordinary steel) has a lower limit of –20°F (–29°C) and an upper limit of 1000°F (538°C); a specific type of Inconel™ (a high-nickel steel alloy made by the International Nickel Company) has limits of

–325°F (–198°C) and 1200°F (649°C). The requirements have been systematized into ANSI standard pressure-temperature ratings. For example, control valve bodies come in standard Classes 125, 150, 250, 300, 600, 900, 1500, 2500, and 4500. Each rating has a maximum allowable working pressure equal to the rating for a specific metal at a specific service temperature. At higher or lower temperatures, the allowable working pressures

(a) Double-acting

(b) Single-acting

Figure 5-8. On-Off Operation of Piston Actuators

vary. These details are covered by tables and charts that are available. To specify the proper class rating for a valve, we determine the extreme combinations of pressure and temperature to which the valve may be subjected, and we select the lowest class rating that is safe for the worst combination. This rating will be the lowest in cost that is technically acceptable, except that we may choose a higher rating if corrosion or erosion of the body wall of the valve by the process fluid may later weaken the body. The standard pressure-temperature classes do not apply to non-metallic bodies. For these, the valve manufacturer should be consulted.

(2) *External forces* that the body is subjected to because of temperature changes of the piping or equipment to which the valve is connected. These forces may distort the valve body by compression, expansion, bending, or twisting. The result may be to damage the valve or interfere with its operation by changing the clearances of internal parts. However, proper design of the piping by piping designers, including general adherence to good piping practices and the valve manufacturer's installation instructions protect against the external forces.

Commonly used body materials are brass, bronze, cast iron, carbon steel, and stainless steel. There are other, more exotic alloy materials.

The valve trim needs adequate strength to withstand fluid forces and the actuator force, but the valve manufacturer normally has the responsibility to provide the strength by using appropriate trim materials.

● *Corrosion.* Process plants frequently use fluids that are very corrosive. Even water, as we know, is corrosive to iron and steel. If the wrong material of construction is used for a control valve, the wetted parts of the valve can be eaten away and become useless or dangerous. Estimates of the annual cost of corrosion to industry are enormous.

The subject of materials is both a science and an art. Tremendous effort has been applied to laboratory and theoretical studies of metallurgy so that the chemistry of metal corrosion may be understood better and so that better metals and metal alloys may be produced. Numberless field tests and experiments have been run. Plastics and ceramics have been studied and tried. Through theory and practical experience, the result has been to improve both the understanding and application of materials for control valves, as well as for other instruments and equipment. Where corrosion is bad but the service is not otherwise severe, a valve body and trim may be of plastic, or some parts may be of plastic-lined metal. Ceramics, too, have been used.

Yet the field results often fall short of those expected because of a lack of sufficient knowledge for a specific application; the subject can be tricky. A very small amount of an unexpected or ignored chemical contaminant may make a great difference in the corrosiveness of a chemical solution.

Sometimes a decision is made knowingly to forgo the use of the best material because of high cost; a compromise is made. Or the best material to resist corrosion may be too soft or have some other undesirable property.

● *Erosion.* Erosion is another complicating factor. Sharp stones in a brook eventually become rounded by erosion caused by the gentle flow of water. The Appalachian Mountains were once as high and rugged as the younger Rocky Mountains are now, but they gradually eroded before the Rockies arose.

The same type of effect occurs in control valves except in a greatly speeded-up way where fluid velocities are high. We have seen that constricting a flow, as is done in a modulating valve, increases the velocity locally. The increased velocity aggravates the erosive action of the stream.

In the case of liquids, increasing the velocity may lower the liquid pressure sufficiently to create localized *flashing*, which means boiling (see Fig. 3-17). If this happens, the resulting vapor bubbles increase the flowing volume within the valve and the velocity increases further. There is then a mixture of liquid and bubbles of vapor zipping through the valve, with an intensified erosive effect. Flashing is damaging to the valve and often to the downstream piping.

There is more. We have seen how a liquid downstream of a constriction recovers some of its pressure. If the piping downstream of the control valve is at a pressure above the vapor pressure (which is a measure of the tendency to boil at the existing temperature), then the vapor in the liquid-vapor mixture condenses and the bubbles collapse. This collapsing, known as *imploding*, causes powerful shock waves that tear at the valve materials and the nearby downstream piping. The phenomenon of flashing followed by implosion is called *cavitation*. If the downstream pressure is low enough to avoid creating implosions, then there is only flashing and the downstream flow remains as a mixture of liquid and vapor.

Cavitation is especially damaging. In extreme cases, within a few hours' operating time it will cause failure of some control valve components that are made of hard alloys. Cavitation is somewhat like sandblasting. It creates vibration and a sound like that of flowing gravel, and it pits metal badly. It is a problem not just for valves but also for pumps and ship propellers, which also create localized increases and decreases of liquid velocity.

The effects of cavitation may be minimized by choosing special types of valve bodies, including the *drag valve* and the *multistage valve*. A drag valve lowers the fluid pressure by passing the fluid through many narrow tortuous passages in parallel. A multistage valve takes the valve pressure drop in a number of steps in series. Proper choice of the piping materials and design of the layout may help protect the downstream piping against cavitation damage.

To minimize damage from all types of erosion, very tough and hard materials are required, especially for the valve trim. An example of hard trim is Stellite (Trademark, International Nickel Company). Where flashing and cavitation are expected, alloys other than ordinary carbon steel are required for the valve body, sometimes even for a short length of piping downstream of the valve.

5-1-6. Noise

Control valves can create audible noise that is sometimes unbearably loud. The fluid velocities in valves may be extremely high, and the amount of power dissipated by a valve can be very large. For example, consider a control valve handling 2000 gpm of cold water with a pressure drop of 1000 psi; this is not an extreme case. The hydraulic power to be dissipated is 1167 horsepower, which creates a loud and harmful noise.

To protect people who work in industrial environments, the Occupational Safety and Health Administration (OSHA) of the federal government has established standards for the maximum allowable noise. For example, a person is permitted to be exposed for eight hours a day to a sound level equivalent to that of an unmuffled truck engine. The permissible level of sound at the source increases as the person moves away from the source and as his exposure time is shortened.

Control valve noise is produced by the following:

(1) The fluid flow.
(2) Mechanical rattling, due to clearance between parts of the valve.
(3) Natural frequency vibration, which is what a flagpole undergoes when there is a strong wind. The valve becomes a sort of tuning fork.
(4) Valve position instability with resulting oscillation.

The control valve manufacturers can provide estimates of the noise their valves will produce in a given application. Proper selection of valve design, valve size, and actuator can reduce the noise.

To reduce the propagation of the source noise, the sound path has to be treated. This may involve the use of any or all of the following: heavy-wall pipe, acoustic insulation, and diffusers or mufflers.

5-2. REGULATORS

A *regulator* is a device that senses a process variable and automatically performs *all* the corrective actions that are needed to keep the value of the process variable nominally at a set point. The device does not require a separate, outside power supply. In other words, it is a self-actuated closed control loop.

Regulators typically have only a non-adjustable proportional control mode, meaning that they cannot be tuned. Generally, they have a set point adjustment. If more sophisticated control is required, then a regular controller and control valve are used. However, regulators fulfill many needs very satisfactorily and, in the smaller sizes, are generally less costly than a controller/control valve combination.

Examples of regulators for pressure, temperature, level, and flow are described below.

5-2-1. Pressure Regulators

Figure 5-9 shows the basic designs of a pressure regulator, of which there are two basic types: the *reducing valve* and the *backpressure valve*, which control downstream and upstream pressures, respectively. In the figure, both types have the controlled pressure increase as the downward force of the spring is increased; manually adjusting the spring adjusts the set point. Opposing the spring force is the upward force created by the controlled pressure acting upon the diaphragm.

These regulators are available for controlling gage pressure, absolute pressure, and differential pressure, depending on whether and how the diaphragm chamber of the actuator housing is pressurized (see Section 3-4-3).

5-2-2. Temperature Regulators

A control valve having a diaphragm actuator, as in Figure 5-6(a) and (b), uses a variable, controlled air pressure to modulate the valve. For a temperature regulator, the air pressure of the control valve is replaced by the pressure from a filled thermal system (see Section 3-7-2). The bulb of the filled system is connected through tubing to the valve actuator. As the sensed temperature increases, the pressure of the filled system increases, thereby moving the valve stem. The diaphragm force from the filled system is opposed by a spring. The arrangement is illustrated in Figure 5-10.

Figure 5-9. Basic Designs of Pressure Regulators

Figure 5-10. Basic Design of a Temperature Regulator

5-2-3. Level Regulators

The emptying of a toilet water tank is initiated manually. The tank has a level regulator consisting of a ball float that operates a water inlet valve to refill the tank automatically and then shut off.

A typical industrial level regulator has a ball float that rides up and down with liquid level changes in a tank. Through a mechanical linkage, the float movement causes a valve plug to modulate the flow to, or from, the tank. The regulator consists of the float, linkage, and valve illustrated in Figure 5-11.

5-2-4. Flow Regulators

Flow regulators typically consist of a differential-pressure regulator combined in one assembly with a valve, which is simply an adjustable orifice. The basic principle of operation is diagrammed in Figure 5-12.

For a fluid at a fixed temperature and pressure, the flow through an orifice is constant if the pressure drop across the orifice is constant. For example, if we open our kitchen faucet, the water flow is constant provided that the city water supply pressure does not change. In Figure 5-12, the differential-pressure regulator maintains a constant pressure drop across the valve by modulating open or closed if either the inlet pressure or the outlet pressure changes; thereby the flow is kept constant. To increase or decrease the flow, the flow set point is changed by opening or closing the hand valve as required.

Figure 5-11. Basic Design of a Level Regulator

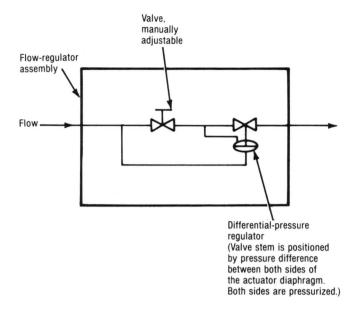

Figure 5-12. Basic Concept of a Flow Regulator

Figure 5-13. A Pressure Relief Valve

5-3. LIMITING ELEMENTS

A *limiting element* is a device that senses a process variable and that automatically takes *all* the corrective actions needed to prevent the value of the process variable from exceeding a set limit. Different elements are suitable for repetitive or for one-time-only operation. Examples of limiting elements for pressure, temperature, and flow are described below.

5-3-1. Pressure-Limiting Elements

• *Pressure Relief Valves.* Safety valves, relief valves, and safety-relief valves are slightly different from each other, but their basic principle is the same. We will refer to them here simply as *pressure relief valves.*

These valves are used to prevent process equipment — piping, vessels, pumps, etc. — from bursting because of an excessive difference between the inside and outside pressures. If a tank is overpressured beyond its design rating, it bursts outward; it explodes. If a tank is under vacuum and has not been designed for it, it bursts inward; it implodes. A valve to prevent implosion is termed a *vacuum relief valve.* A typical relief valve to prevent explosion is illustrated in Figure 5-13.

The inlet of the pressure relief valve is piped to the equipment or system that is to be protected. In normal plant operation, the spring pushes the valve plug down against the process pressure, and the valve is closed. If all goes well, the valve will never be required to open during the life of the plant. But if the pressure were to become abnormally high because of control system malfunction, operator error, equipment failure, external fire, or chemical reaction, the valve would then open to limit the pressure buildup. This occurs when the process pressure against the bottom of the valve plug creates a force that exceeds the downward force of the spring at the spring set point plus the weight of the disk assembly.

The spring force is adjustable to change the valve set point.

For the valve to open fully, the pressure must rise some predetermined amount above the set point; this incremental amount is known as *accumulation,* which is expressed as a percentage of the set point — typically 3 percent, 10 percent, or other, depending on the application. A properly sized pressure relief valve, when wide open, limits the rise of pressure (or vacuum) to an acceptable value for safety.

A pressure relief valve is suitable for repetitive operation though it tends to develop leaks when it is operated, especially at high pressures.

Mathematical formulas are used for sizing the valves according to the service conditions.

There are stringent standards established by ASME, the American Petroleum Institute (API), and other engineering bodies regarding the design of equipment for pressure or vacuum service and the over- or under-pressure protection requirements. These standards in many cases have quasilegal status through endorsement by governmental bodies.

The most widely used code for this subject in the United States is the ASME *Boiler and Pressure Vessel Code* [64], which covers the requirements for pressure relief valves. Many pressure relief valve manufacturers produce valves that bear an ASME nameplate attesting that the valves meet the ASME requirements. However, pressure relief valves are used wherever needed for safety, regardless of whether there is an explicit code requirement for the valve.

Both relief valves and backpressure regulators limit a rising pressure (see Section 5-2-1). Some pressure relief valves open proportionally with increasing pressure, as do many regulators; others snap open. However, unlike pressure regulators, pressure relief valves are intended only to limit pressure, not control it.

The pressure relief valve codes are an essential point of difference between pressure relief valves and pressure regulators. The ASME code requirements can be met by pressure relief valves but generally not by regulators, and generally not by an ordinary pressure control loop.

Good practice generally has at least two parts to ensure pressure safety, as follows:

(1) The first part consists of ordinary instruments to keep the pressure within safe bounds during plant operation. This part may be a pressure reducing valve or a control loop, which controls the inflow to the process system, or a backpressure valve or control loop, which controls the outflow.

(2) The second part normally consists of a pressure relief valve (or a rupture disk — see below) that should never have to perform if the first part does its job, but that stands ready to perform if the first part fails. This practice follows the principle of redundancy (see Section 10-2-3).

● *Rupture Disks.* Rupture disks, also known as *safety heads*, are recognized by the ASME *Boiler and Pressure Vessel Code*. Like pressure relief valves, they are used to protect equipment against excessive pressure or vacuum. They are used either in place of or together with a pressure relief valve, depending on the application.

A disk is a thin diaphragm or plate that is usually mounted between a pair of pipe flanges and closes off the pipe. The disk may be made from a variety of metals, graphite, or other materials. Metal disks are sometimes plastic coated to protect the disk against thinning and premature rupture from chemical attack. Figure 5-14 shows a typical rupture disk assembly.

The operation of a rupture disk is simple: When the process pressure rises to the design bursting pressure of the disk, the disk bursts, splits wide open, and the pressure is relieved.

The design bursting pressure is equivalent to a set point but is not adjustable. Its value depends on the disk being properly manufactured and is, therefore, not very accurate. In addition, metal disks may fail prematurely from metal fatigue. The lack of accuracy must be taken into account in designing the process equipment and the associated control system.

Nevertheless, rupture disks are very useful, especially when dealing with corrosive fluids or when all leakage from relief valves is to be avoided.

A ruptured disk is not reusable. Disks may be made of expensive metals that have salvage value.

Figure 5-14. A Rupture Disk Installation for Pressure Service

5-3-2. Temperature-Limiting Elements

Fusible plugs or *fusible links* are temperature-sensitive devices that are used to limit temperature. A fusible plug may be a plug that normally blocks the outlet opening of a spray nozzle of a water sprinkler intended to fight a fire. Or a fusible part may be in the form of a mechanical link that normally holds open a fire door that should be closed in case of fire. The plug or link is made from special metal alloys that melt at relatively low temperatures.

If the fusible part is in a location whose temperature becomes abnormally high, the part is destroyed by melting. The water flows through the spray nozzle, and the fire door is free to close by rolling down an inclined support.

A fusible part senses the process variable, temperature. The temperature set point is built in through selection of an alloy having the proper melting point. The control action is to disintegrate. Once used, the fusible element must be replaced.

5-3-3. Flow-Limiting Elements

An *excess-flow check valve* is a valve that permits a fluid to flow until the flow rate becomes excessively high, at which time the valve stops the flow. The cutoff flow is typically 1-1/2 to 2 times the normal flow. An excess-flow check valve is different from an ordinary check valve, which does not block forward flow but does block reverse flow. Figure 5-15 shows the basic design of an excess-flow check valve.

Flow ⟶

Figure 5-15. Basic Design of an Excess-Flow Check Valve

Reverse flow tends to close the valve but a spring tends to keep the valve open. The spring is adjustable and provides the limit set point. As the flow increases, the closing force increases and, at some point, the spring force is overcome and the valve closes.

Excess-flow check valves are used for safety against excessive flow caused by a broken line or other mishap. They can be used repetitively.

5-4. OTHER FINAL CONTROL ELEMENTS

In previous sections, we saw how control systems use control valves as the final control elements to manipulate process variables. Other kinds of final control elements are presented below.

5-4-1. Proportioning Pump

A *proportioning pump*, also known as a *metering pump*, pumps liquids by means of a piston or plunger inside a cylinder. Every back-and-forth stroke cycle of the pump discharges a fixed volume of liquid. The length of stroke of the pump can be adjusted manually, or remotely by the use of a pneumatic or other type of actuator. Each such adjustment changes the discharge volume per cycle.

Proportioning pumps are usually of low capacity and are frequently used to inject small volumes of a chemical solution into a relatively large stream. An example of such use is the chemical treatment of boiler feedwater in a power plant to protect the steam generator against impurities in the feedwater. A typical system uses a dissolved-oxygen analyzer, a controller, and a proportioning pump to inject hydrazine (a chemical) to lower the concentration of oxygen dissolved in the water. The pump is the final control element of a modulating analysis control loop. This is shown in simplified form in Figure 5-16.

5-4-2. Vane Control of Fans

Industrial fans may be supplied with adjustable *vanes*, or *dampers*, or *louvers* for manipulating the gas flow through the fan. The vanes are placed on the inlet side of the fan and act, in effect, as a control valve. They can be actuated either automatically as part of a control loop or manually.

This diagram shows the basic concept for injecting a chemical, hydrazine, to reduce the dissolved-oxygen concentration in the feedwater to a steam generator.

Figure 5-16. Using a Proportioning Pump as a Final Control Element

5-4-3. *Variable-Speed Control*

The flow of materials can be modulated by varying the speed of process equipment that uses variable-speed drives. Examples are:

(1) Conveying solids by means of a variable-speed screw conveyor or conveyor belt.

(2) Transporting liquids or gases in a pipe or duct by means of a variable-speed pump or compressor, respectively.

Power drives offering variable speed for process control include:

(1) Electric motors.
(2) Pneumatic motors.
(3) Hydraulic motors.
(4) Mechanical transmissions.
(5) Steam turbines.

Variable-speed drives can be costly to purchase and maintain. The benefit they provide is to reduce power consumption by using no more power than is needed for pumping at any given time.

6

How Logic Control Is Performed

6-1. DEFINITION OF LOGIC CONTROL

Logic control is control that causes things to be done or not done to the process, depending on whether certain process conditions, operator actions, or control system actions happen or do not happen. Logic control makes use of a logic system, which is a combination of binary elements that automatically "decide" on the proper response to a certain set of conditions. The concept of logic control is outlined in Figure 6-1.

In logic control, every part of the logic system — each input event, each logic function element, and each output response — is either YES or NO, ON or OFF, TRUE or FALSE, 1 or 0, as we saw in Section 3-2-2 for binary signals and in Section 4-2-1 for binary control. Logic control is appropriate for any process system that uses on-off devices to initiate or terminate normal or emergency operations.

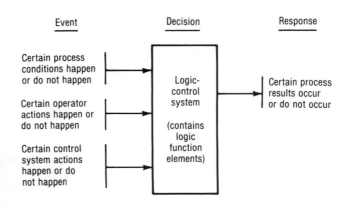

Figure 6-1. The Concept of Logic Control

6-2. HOW LOGIC CONTROL WORKS

Let's bring the definition in Section 6-1 down to earth. Some people we know are planning a picnic. Let's see how they do it.

Abner and Bella want to go on a picnic with Chauncey and Desiree. If they cannot all go this weekend, then they will go the following weekend. The process to be carried out is to get ready for the picnic and go. Figure 6-2 diagrams some of the possibilities. The symbolism used is taken from ANSI/ISA Standard S5.2-1976 (R1981), *Binary Logic Diagrams for Process Operations.* [56]

The figure is a logic diagram that shows the following requirements that need to be satisfied in order for the group to go on the picnic this weekend:

(1) Nobody is sick.
(2) Either (a) Abner does not get an emergency call to work on this weekend, or (b) he gets the call but is able to get Zorro to fill in for him.
(3) Food and drink are prepared.
(4) A car is available.
(5) The weather forecast is good.

Figure 6-2. Getting Ready for a Picnic

If every one of these requirements is met, then the group goes on the picnic this weekend. If a single one or more of the requirements is not met, then they must wait a week.

The AND function in the figure requires all its inputs to exist and continue to exist in order to obtain and maintain a logic output; if any input ceases, the AND output ceases. The OR function requires only one or more of its logic inputs to exist in order to have and maintain a logic output. The circle represents the NOT function, which really means IF NOT: If there is no logic input, then there is a logic output, and if there is a logic input, then there is no logic output. The logic functions are sometimes called *gates*; for example, an *AND gate.*

In the above example, the logic signals and function symbols exist only in our minds. But for a process plant, a logic diagram uses actual hardware to carry out the logic, as described below.

Consider a process plant situation: The plant has a chemical pump that is started by push button. The pump requires a supply of lubricating oil before it starts. The supply could be turned on manually but, for reliability, it is decided to do it automatically. So the following must first occur: A lubricating-oil pump must start automatically when the chemical pump START push button is operated, and normal oil pressure must be established within 6 seconds. The logic sequence of operations is as follows:

(1) If the oil pressure is normal — that is, not low — after 6 seconds the chemical pump operates. Otherwise the startup is cancelled automatically.
(2) The operator can stop the operation at any time by pressing the STOP push button. Independently, the operation stops if the oil pressure becomes low.
(3) When the chemical pump is commanded to stop, its power is cut off immediately and it stops, but the oil pump should continue to operate for 10 seconds, then stop. The 10-second delay ensures that the chemical pump continues to be lubricated while it is slowing down.
(4) While the chemical pump is required to operate, an alarm, PAL-64, is actuated if the oil pressure becomes low.

Figure 6-3(a) diagrams the logic for starting and stopping the chemical pump normally and for protecting the pump against low oil pressure during operation. We should remember that the operating lines in the figure represent logic signals, not physical signals.

Now consider the case in which the system is down and there is no oil pressure, which is a normal situation

(Each pump stops operating when the logic does not call for it to operate.)

(Note: This arrangement permits a nuisance low-pressure alarm. See (b))

All information moves to right or down, unless otherwise noted.

(a) Basic logic

(Note: The input for the alarm is relocated away from the source shown in (a)).

(b) Logic with nuisance alarm eliminated

Legend:

A   = AND function
DI = *Delay Initiation* of logic output for a specified time
DT = *Delay Termination* of logic output for a specified time
HS = Hand switch
OR = OR function
PSL = Low-pressure switch
PAL = Low-pressure alarm

◯ = NOT function (meaning IF NOT)

S
R = A memory that is established when set and persists until it is reset:
     S = *Set* memory
     R = *Reset* memory

Figure 6-3. Operating Logic for a Chemical Pump

when the chemical pump is not required to operate. A low-oil-pressure alarm is triggered, which is false because there is no trouble. Nevertheless, the alarm requires an acknowledgment action by the operator, and the alarm light remains on because the seeming trouble — low oil pressure — persists.

Such a false alarm is known as a *nuisance alarm*, which exists here because the system logic does not recognize that the system is in a shutdown state. Nuisance alarms are an inconvenience and, worse, a distraction for the operator (see "Presentation of Information" in Section 10-3).

A nuisance alarm can be eliminated by permitting the alarm to be actuated only in case of a genuine trouble or abnormality. In our example, the alarm logic should be as follows, with an added anti-nuisance requirement shown in italics: While the chemical pump is required to operate, an alarm is actuated if the oil pressure becomes low *provided that 6 seconds have elapsed after the START button is pressed.* Thus, the oil pump operates and has time to build up the oil pressure before the alarm can operate.

Figure 6-3(b) shows the logic for the pump system with the nuisance alarm blocked.

All the stated logic conditions and actions are binary in nature: The oil pressure is either low or not low, signalled by the output from the on-off switch, PSL-64; the time-delay elements have an output or not, depending on whether the specified time has expired or not; the pumps operate or not; and so on.

A different situation may exist where there is an analog control loop whose action should be blocked under some circumstance. For example, Figure 6-4(a) shows a modulating flow control loop controlling the flow of liquid to a tank. The outflow from the tank is controlled by a level controller. No excessively high level is expected, but if it were to occur through some mishap, the flow to the tank must stop.

The flow controller signal passes through a small, solenoid-actuated, three-way pilot valve that normally does not hinder the signal to the control valve. The controller signal is pneumatic, 3 to 15 psig. But if the tank level becomes abnormally high, the high-level switch output shifts from ON to OFF, the pilot valve moves to vent the control valve actuator, the control valve fails closed on loss of control air, and the flow to the tank is stopped.

What we have here is a binary control loop — the level switch and its pilot valve — that overrides an analog controller. The basic logic is shown in Figure 6-4(b). The override, also known as an *interlock,* disrupts the process operation by stopping the flow to the tank. Compare this with the selective control of Section 4-3-3, which has a mutual override of two analog controllers with no interruption of the process. Each scheme has its place.

A kitchen example of a safety interlock is a microwave oven that cannot operate until the door is closed.

6-3. THE ADVANTAGES OF LOGIC CONTROL DIAGRAMMING

Logic diagramming arose in response to a need for planning logic systems that were far more complex than the system of Figure 6-3. The parts of written or spoken verbal descriptions of interlocking process operations are often difficult to piece together mentally to provide a whole picture. Logic gaps or errors may go undetected.

The logic control plan should be correct and clear before we attempt to detail the hardware circuits to carry out the plan. Otherwise, we risk getting on a horse and charging, so to speak, before we know where we are going. Let's do it right the first time, the zero-defect way.

The few symbols of the logic language together with some English language words enable a coherent and explicit description of the operating logic to be diagrammed. After a little familiarization, the graphic nature of the diagram makes the diagram relatively easy to read and understand; logic inconsistencies become easier to catch. The method is simple in that it focuses on the process operating requirements, not the details of any particular kind of hardware. Therefore, the logic depends, for example, on whether a flow is high or low, not on whether the electrical contacts of the flow switch are open or closed. The mechanical details of the switch action are left for the technical specialist.

At the same time, the method can present the operating requirements with a greater or lesser degree of detail depending on whom the diagram is intended for. The level of detail may range from that required for a designer of detailed control circuits to that for a project engineer who needs some understanding of the operation of the plant that his subordinates are working on. Other types of people who may get involved include process engineers, equipment engineers, plant operators, and service personnel.

The electrical schematic or equivalent circuit diagram is the definitive diagram to show how the operating

(a) Overall scheme

(b) Basic binary logic

Figure 6-4. Analog Control with a Binary Interlock

requirements have been carried out. It reflects the circuit designer's knowledge of the intricacies and idiosyncrasies of control circuits and their hardware. Incompatibilities that may exist between the logic diagram and the actual circuit must be ironed out before either diagram can be considered final.

6-4. IMPLEMENTATION OF LOGIC CONTROL

6-4-1. Logic Circuit Components

All logic control functions can be carried out using several different types of hardware, unmixed or mixed. The hardware consists of individual components that are connected together by wire conductors, printed electrical circuits, tubing, or optical fibers to form a logic control system, as in Figure 6-1. The components of a system may include the following types:

- *Electromechanical Relays.* These relays are of the traditional electromechanical types consisting of an electrical solenoid that, when energized or de-energized, mechanically moves one or more electrical switch contacts to open or close electrical circuits.

- *Electrical Solid-State Devices.* These arose following the invention of the electrical transistor. They perform switching operations but have no mechanical movement. They are steadily supplanting electromechanical relays as control elements but, generally, have a smaller current-carrying capacity.

- *Fluid-Power Devices.* These are usually pneumomechanical devices, essentially very small air-actuated valves and power pistons. They are very widely used for controlling machine operations in manufacturing plants. Pneumomechanical devices have limited application in process plants, usually in small, field-mounted systems; the larger systems are normally electrical.

- *Other.* These include fluidic devices, which use air but have no moving parts; hydromechanical devices; optical devices; and manual devices.

6-4-2. Programmable Logic Controllers

For a programmable logic controller, the designation *PLC* is preferred to *PC*, which is widely used to designate a *personal computer*. (PLC is a convenience term that is not related to the instrument identification scheme of ANSI/ISA S5.1. [1])

A PLC is a programmable electronic instrument that has a video screen and a manual keyboard. *To program* the instrument is to enter into it instructions for solving a problem. Originally the problem to be solved was only to determine and provide the proper binary output or outputs according to the logic of the binary inputs, but the scope of the problem solving has been greatly enlarged, as described below. The programming is performed by using the keyboard. The program results are shown on the screen.

A *programmable logic controller* is really a *configurable logic controller* because it has already been programmed by the manufacturer. It merely needs to be configured by a person choosing among the program options that the manufacturer has provided; the ease of configuring is an outstanding benefit of a PLC. (See the discussion of configuring under "Computer Control" in Section 7-3-2.)

After the process control logic is established by means of a logic diagram or the equivalent, an engineering diagram is required if electromechanical relays will be used. Such a diagram is called a *relay ladder diagram*, simply *ladder diagram*, or *electrical schematic diagram*. (See Figure 6-5.)

PLCs are digital computers (see the discussion of direct digital control in Section 7-3-2) that can now perform the following functions [36]:

(1) Logic.
(2) Sequencing.
(3) Timing.
(4) Counting.
(5) Calculating.

Drawn manually

Free-format equivalent PLC diagram

Legend:
1PB	= Push button
2CR, 4CR, 5CR	= Control relay switch contacts
3LS	= Level switch
SOL A	= Solenoid-actuated device

Figure 6-5. A Relay Ladder Diagram

(6) Analog control, including the proportional, integral, and derivative modes.
(7) Working with other PLCs in a network.
(8) Working in a direct digital control network (see Section 7-3-2).

The Company Hierarchy Of Control

7-1. ACHIEVING THE COMPANY'S PRIMARY GOAL

So far we have focused on the plant process: how to measure what is going on and how to control it. However, the plant that is supposed to operate efficiently is only one element of a grander operating system of a different kind. The larger system is the plant owner's whole organization, whose overall purpose is to run a profitable business by satisfying a public demand for the plant product. This is why the owner invested his effort, time, and money in establishing the plant. The production plant is only a means to an end.

For the company, the profit-making process is the production and sale of goods, which is part of a closed-loop control system, as shown in Figure 7-1. This loop is an application of the generalized closed control loop of Figure 4-1. The profit goal is the control set point, which really is a minimum goal: The owner does not object to overshooting the goal, but he would rather not fall short.

The production portion of the process involves obtaining and using information concerning the manu-

facturing process. Some of this information — on production, inventories, and so forth — is used also by upper-level managers for planning.

7-2. THE CONTROL HIERARCHY

Company control of plant operation has several layers of responsibility, a hierarchy of control. There are no hard-and-fast categories of control that fit the needs of different plants and different companies. However, a representative arrangement may be as shown in Figure 7-2. The figure outlines the general plan and information relationships for three ranks of authority and responsibility in a company having several production plants.

The ranks of control, beginning with the highest, are as follows (also see *distributed control* in Section 7-3):

● *Rank 3* is the central management, which plans the master strategy for the company operations; its people may be likened to the general staff in the army.

The management is concerned about the balance between production and sales, among other things. For example, the management of a petroleum refining company is informed that the coming winter is expected to be unusually cold. Their planners decide that they want to produce more heating oil and less gasoline. Based on analyses of the performance, capabilities, and inventories for each of their oil refineries, the planners specify the product mix they would like to see from each refin-

ery in order to optimize the performance of the company as a whole. The responsibility of each plant for meeting the production target falls to Rank 2.

● *Rank 2* covers the plant management for each refinery. The management plans the strategy to optimize the operations so that the refinery may meet its production goals most efficiently; they establish the operating requirements for the refinery. This management is like the commissioned officers of the army who plan the

A

Company Departments

Accounting
Development and design
Marketing
Personnel
Plant administration
Purchasing
Quality control

B

Manipulated Variables

Manufactured product
— Quantity
— Design
— Quality
Materials
— Type
— Cost
Management and personnel
— Quality
— Remuneration
Production facilities
— Automation
— Equipment quality
— Working conditions
— Number of plants
— Plant locations
Marketing
— Commercial arrangements
— Product packaging
— Advertising
Income and expenses
— Selling price of product
— Fixed costs
— Variable costs
— Financing
— Cash flow
Political lobbying

C

Disturbances

Owner's and competitor's changes
　in variables of column *B*
Public taste
Improved technology
Labor disturbances
Loss of materials sources
Public unemployment
Natural phenomena
War
Government
— Labor regulations
— Mandated employee benefits
— Environment regulations
— Domestic subsidies
— Export subsidies
— Import restrictions
— Export restrictions
— Taxes
— Interest rates
— Monetary inflation or deflation

Figure 7-1. Controlling Company Profit

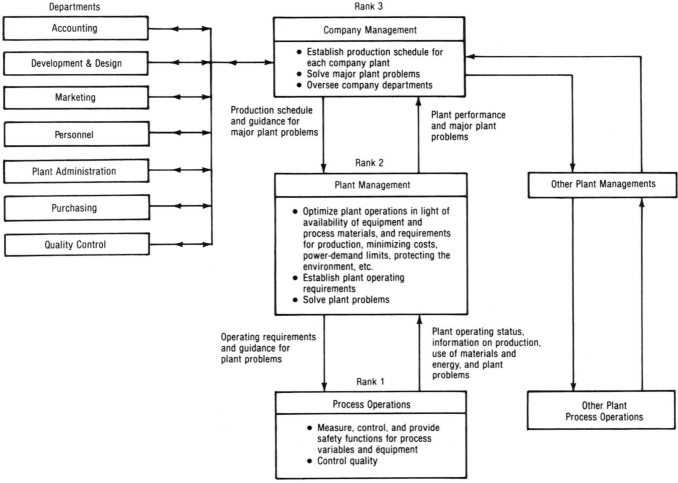

Arrows show the flow of information.

Figure 7-2. The Company Hierarchy of Control

tactics to achieve the military goal they have been given by Rank 3.

• *Rank 1* consists of the operating people who directly control the process in their particular plant. These people are like front-line junior officers and enlisted men; they carry out the tactics for victory as directed by Rank 2.

Rank 1 makes use of process instrument systems for measurement and control. The control systems may include the following types, which are discussed in Section 7-3:

   (1) Traditional field control
   (2) Traditional centralized control
   (3) Computer control, leading to:
      (a) Supervisory control
      (b) Direct digital control
      (c) Distributed control

7-3. THE FORMAT OF PROCESS CONTROL LOOPS

The measurement and control systems that are used by Rank 1 for direct control of plant operations have different formats, that is, different arrangements, for separating or grouping loop or instrument functions in different plant locations. The word *architecture* may be used instead of *format*. The format of the systems is outlined below. However, the field of instrumentation is in flux, with constant mutation and hybridization of basic instruments and their systems. It is difficult to cover the field broadly without finding exceptions to some statements in the descriptions. However, the descriptions do point out practices and possibilities.

7-3-1. *Traditional Instrumentation*

• *Field Control.* The domain of process instrumentation began with only field control, which remains important now but not exclusively so. Furthermore, field control will continue to be used for many services in the future.

Field control has all the instruments of a loop locally mounted, generally near the related plant equipment. Field control provides operating information right at hand for operators who are stationed there or who make a routine once-a-day or once-a-shift inspection of the plant equipment and operation, and also for people who service the equipment.

Field control is also the simplest system. It has the fewest links to break in the instrument chain, thus enhancing system reliability. It avoids the high cost of running signals from a local point in the plant to a remote control center.

Such control is perfectly appropriate today in many cases for control loops that have a constant set point, as is often true for plant utility systems. For example, a plant produces steam at 300 psig but requires some steam to be used at 15 psig. A local pressure-control loop that is set unchangingly to control at 15 psig will often be the correct type to use; there is no need for operator adjustment, hence no need for the controller to be in a control center for frequent manipulation. (There may be a remote alarm system to guard against the consequences of a failure of the local control system. The very earliest alarms were also local.) Figure 7-3 illustrates the arrangement for direct-connected instruments, showing both how they are used and the system format. The control is totally local and requires a local operator to carry out the Rank 1 responsibility. Ranks 2 and 3 have no tie-in to Rank 1 other than through the classical means of written or oral communication.

• *Centralized Control.* The locations of local control loops in the traditional field arrangement were widely scattered. In order to reduce manpower requirements and improve the coordination of operations of the various parts of the plant, instruments became less isolated, beginning with local groupings and culminating in a control room that now is the control center for the entire plant. The control room is designed so that all important operations can be given close surveillance and be carried out from this location.

The centralized instruments were originally mounted in rows on instrument boards of steel or other material. The boards were at first made of vertical panels, approximately 7 feet high, placed side by side in a straight line. Then the boards became U-shaped because of their great length, up to 50 feet or more, and the vertical

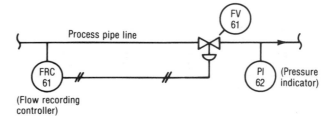

(a) Process application

Location	Function	Instrument
Field	Set point	FRC
	Control	FRC
	Readout	FRC, PI
	Manipulate	FV

(b) Loop format

Figure 7-3. Traditional Field Control

panels developed sloping segments at the top and at bench height for the convenience of the operator and to improve the readability of the instruments. (It was claimed but never authenticated that the operators needed roller skates to cover some of the large boards.)

Instruments used to be large, approximately 15 inches, or more, in height and width. In the late 1940s, miniature indicating and recording controllers, 6 inches wide by 6 inches high — later in widths down to one inch — were introduced and permitted closer spacing of instruments and the placing of a greater number of instruments on the central board. These dimension changes took place initially for pneumatic instruments, then for electronic instruments. The concentrating of instruments was accompanied by the replacing of relatively bulky alarm lights by annunciator cabinets with windows that utilized board space more efficiently.

The additional instruments on the board increased the amount of process information that the operator had to deal with. At the same time, the miniaturization of instruments helped to make practical the graphic panel, which was often used to improve the operator's command of the process. A *graphic panel*, sometimes called a *mimic panel*, is a panel bearing a simplified diagram of a plant process or a portion of it. The diagram is usually made from flat plastic pieces; color coding is used to make the diagram easier to follow. The control instruments, readout instruments, and lights are placed in the diagram where they fit in the process.

ANSI/ISA Standard S5.5-1985, *Graphic Symbols for Process Displays*, has established standard graphic symbols for use in process displays on both graphic panels and video screens. [74]

Another development to help the operator was the control *console*, which was a low instrument board with a projecting shelf. The control console was to be approximately U-shaped, patterned after an organ console. The theory was that a seated virtuoso operator, by masterful and deft movements of his arms and fingers, would be able to reach and operate all the instruments needed to make the process plant play in harmony (even without the benefit of bass pedals). An air pilot's cockpit has also been fancied as the prototype for the console.

As consoles developed, they remained faithful to their name but, unfortunately, not always to their intent: An operator would often need an extensible arm or the long arm of a gorilla to perform on his instrument if he were to remain seated. Nevertheless, consoles have eased the operator's job considerably. This is also true for the distributed control systems that have followed.

Data loggers, which are printers of as many as hundreds of pieces of process data, often with alarm, calculation, and simple control capability, began to appear in approximately 1960. In the 1970s, data loggers began to sprout video terminals with keyboards. These were essentially improved devices to do the same old

things but with more flair. They did not cause a basic change in the physical format of control or of control centers.

The gathering together of instruments in the control room requires transmitters to bring the process measurements from the field to the control room instead of having process sensing performed directly by the remote controller or indicator. It is generally more practical, safer, and less expensive to communicate by pneumatic or electric signals over long distances, even within a plant, than to run sensing lines over those distances.

Figure 7-4 shows two variations of traditional centralized control loops. Part (a) of the figure uses a conventional, self-contained controller, which contains two groups of functions that are stated below to illustrate the matter of format:

(1) Control functions: This group includes (a) the control modes — proportional, integral, and derivative — and (b) the configuration adjustments for tuning the control modes and for choosing direct or reverse action. An instrument that contains these functions is a *controller*. It may also contain other functions.

Function	Instrument	Location
Transmit	FT	Field
Manipulate	FV	
Readout	PI	
Set point	FRC	Control room
Control	FRC	
Readout	FRC	

(a) Process control format with a self-contained controller

Function	Instrument	Location
Transmit	FT	Field
Control	FC	
Manipulate	FV	
Readout	PI	
Set point	FRK	Control room
Readout	FRK	

(b) Process control format with a split-architecture controller

An equipment room or the back of the instrument board may be used if additional math or logic functions are required.

Figure 7-4. Traditional Centralized Control

(2) Operator functions: This group (a) indicates the operating variables — process measurement, controller output, and often the deviation from set point and a deviation alarm — and (b) contains the operator's adjustments for set point and selection between automatic and manual control. An instrument that contains the operator functions and does not include the control functions is designated a *control station*.

All controllers used to contain both control and operator functions. With miniature instruments, which initially were pneumatic, came the frequent separation of these functions into what is known as *split architecture*. The purpose of the separation was to permit the control function to be located close to the process in the field while retaining command by the operator in the control room. This was done to eliminate the pneumatic transmission time lags to and from the control room, thereby improving control significantly for fast loops — typically pressure control and flow control.

Most control room controllers today are electronic and have no signal transmission lag. However, split architecture continues to be used, particularly for large plants, for the following reasons:

(1) Relocating the configuration hardware permits the control station to be made smaller, resulting in a more compact instrument board.

(2) The control room operator may not need access to the configuration adjustments, if these are left for an instrument technician to handle elsewhere.

Figure 7-4(b) uses the control scheme of part (a) except with split architecture. If the remote controller is pneumatic, its location is near the control valve; if electronic, the location is in an equipment room or other convenient place.

The format of traditional centralized control fulfills the requirements of Rank 1 and also provides a more comfortable environment for the control room operators. This format tends to reduce the number of plant operators needed and simplifies the job of running the plant as a unit rather than as a collection of disjointed parts.

With Rank 1 having centralized operating information and the capability to gather and report information by means of data loggers and other printing devices, Rank 2 management has easy access to data on key operating variables in the plant and can be kept more up to the minute than in the past regarding plant status. Rank 2 may use a computer to derive information, such as plant efficiency, and to communicate with Rank 3.

Generally, the operation of Rank 3 is not much changed and depends largely on plant information obtained from Rank 2.

7-3-2. The Microprocessor Revolution

A *microprocessor* is an electronic computing element, a fraction of an inch in size, that can contain the equivalent of tens of thousands of electrical circuits and that is still growing in capability. Microprocessors have brought on the Second Industrial Revolution, which is going on in front of us right now.

The revolution began with the invention of the transistor in 1947 by William Shockley, with John Bardeen and Walter Brattain of the Bell Telephone Company. This led to the microprocessor that has transformed commonplace objects — wrist watches, cameras, kitchen ovens, calculators, and innumerable others. In process control, the microprocessor has improved individual instruments and made them more versatile, and it has caused radical changes in control systems. The systems come in almost endless variations, of which representative examples are described below.

• *Computer Control.* The computer has become a major player in the game of process control. The computer was first used as a large machine that simply gathered and recorded data. It was a high-class multipoint recorder (see *data loggers*, under "Chart-Type Recorders" in Section 3-3-2).

The next step was to use the large, or main-frame, computer for *supervisory control*, or *set point control*, in which the computer performed control computations and sent a set point signal to an analog controller, which would do the actual controlling.

The large computer was next used for direct digital control (DDC), which bypassed the intermediate analog controller and operated the final control element directly. The analog controller could be retained as a standby in case the computer broke down. This system had an operating program with the *potential* to model the process, optimize it, and provide non-standard control functions. It seemed like the answer to many problems.

DDC based on the main-frame computer never went far. The computer was a multipurpose machine that needed custom programming to instruct the computer in (1) how to become a controller, and (2) what it should do after it became a controller.

This was comparable to your hiring me and sending me to a school for butlers so that I would learn how a butler behaves and also learn the skills of a butler; I would be programmed to "buttle". After I earned my butler's diploma, you would rent me out to parties. For each party, someone would configure me so that I would know whom to announce on his or her arrival, where the dishes were kept, when dinner was to be served, etc. I would be reconfigured for the next party but my programming would not change.

At that time, the application skills for programming a computer for process control hardly existed. A forced marriage, so to speak, and close cooperation were required between control engineers who did not know how to program a computer and computer programmers who did not know processes or process control — a bad match. In each case, the programmed product was one of a kind; it was expensive, slow to be made ready, and slow to be revised. Another negative factor was that many present-day reliability techniques for electronics and sophisticated control systems were not then established; the risk of shutting down an entire process because of a failed part in the one controlling computer was generally not acceptable.

DDC has progressed so that small, modular, standard controllers are available; however, they are not standardized among manufacturers. A major difference between the old and the new DDCs is that the old required programming and configuration, and the new requires only configuration. The manufacturer now builds in the programs for control modes, calculations, logic, etc., which are called *algorithms* or *firmware*; the controller already knows how to control. All that the controller now lacks is a knowledge of the specific adjustments for each application, and these are readily supplied by an instrument engineer or technician. The user does not require programming knowledge.

One brand of completely self-contained DDC controllers has removable internal function modules that can provide a large choice of control arrangements such as single loop, cascade, ratio, logic, etc.; control modes; calculations; alarm criteria; input signals; and output signals. The controller has analog and digital displays and alarm signals. It can be operated by a computer or by a distributed control system (see below), or it can operate as an ordinary but extremely versatile independent controller. All this capability is contained in a case that is 2.84 inches wide by 5.67 inches high (72 × 144 mm), conforming to an international instrument standard.

DDC controllers are now very reliable; therefore, controller failure is uncommon. The operating domain of each controller is limited; generally, a DDC controller failure does not bring a plant to its knees any more than would a traditional analog controller. Some DDC instruments routinely and automatically test their own circuits and routines.

Another type of computer control makes use of an ordinary personal computer supplied with commercially available expansion boards or interface devices, and appropriate software — the operating programs. Following are typical capabilities, which require use of a printer in some cases:

(1) Working with analog, binary, or digital inputs and outputs.
(2) Data logging.
(3) Strip-chart recording.
(4) Plotting and storing of graphs, curve fitting.
(5) Controlling the process.

With DDC or other computer control, the communications among Ranks 1, 2, and 3 are generally like those utilizing an up-to-date traditional, centralized control system.

• *Distributed Control.* The dispersed instruments of traditional field control constituted a kind of distributed control. Because the instrument systems were generally independent functionally, the failure of one instrument loop did not interfere with the operation of another loop. However, the format had the accompanying disadvantages of compartmentalizing plant operations and requiring considerable manpower.

Traditional centralized control tended to counter these disadvantages but required a large number of costly tubing and wiring runs between individual local instruments and the control room. The consolidating of instruments in the control room steadily increased the information traffic that the operators handled; the effects were as follows:

(1) The operators' job became easier because the operating information and important controls they needed were now close at hand. Also, their vision of plant operations was now more panoramic.

(2) The large flow of information in and out, especially in a fast-changing emergency, sometimes created an overload of information for the operators. They would have to respond to one or more alarms, search for the related system instruments, read them, evaluate the information, decide on the proper emergency action, and finally take action. The consequences of a wrong response can be serious human, physical, and financial damages, as evidenced by the catastrophic Chernobyl nuclear power plant accident in Soviet Russia in 1986.

This led to the next stage of evolution: the present-day distributed control, whose development was begun in 1969. *Distributed control* is a control system having the following dominant features:

(1) Measurement, control, and communication are performed by groups of modules that are distributed in function and location.

(2) The control of the process by the plant operators is performed in a centralized control room.

(3) Local operating control stations may be scattered over the plant.

(4) If control from the control room is lost, the local operating stations continue to function.

(5) If one local operating station fails, all the other local operating stations continue to operate.

(6) A common communication channel, the data highway, runs through the plant and connects to all parts of the distributed control system.

(7) A distributed control system can start small and expand as needs require and circumstances permit.

Figure 7-5 illustrates a representative system, which includes subsystems that are situated in several basic locations as follows:

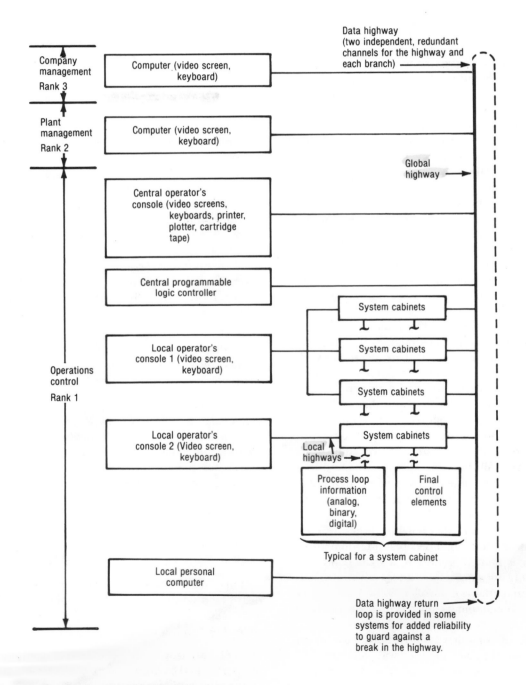

Figure 7-5. A Distributed Control System

(1) *Local.* The system has process measurement instruments that connect directly to local system cabinets. The instruments may be sensors, transmitters, status switches such as a position switch on a valve, or other devices. Other local system cabinets contain elements for control, logic, storage of information, calculation, and so on. (Distributed control has introduced a new minor semantic problem. These local cabinets, unfortunately, are sometimes called *controllers*. They are not the control-mode type of controllers that we discussed previously, and they should not be confused with them, even though they may contain true controllers.) Local final control elements respond to commands from system cabinets. There may be either no local consoles for the operators or one or more consoles for partially centralized local control of sections of the plant. For example, it may be helpful to have some local controls for the pulverizing section and the water-treating section of the plant. The operators' controls and supervisory information are assembled at the local consoles. The information that a local console receives may be displayed on a video screen. The operator may issue commands typically by using a keyboard that is connected to the screen. A local personal computer may be tied into the distributed control system.

(2) *Central Control Room*: Here is the focal point for controlling the operations of a plant area. The room contains a central operator's console, which is intermediate in size between a local operator's console and most traditional central control boards. This console usually has the following:

> (a) Two to six video screens, which can be viewed simultaneously, together with keyboards. These can be used for process control and to bring up information. Having more screens means displaying more and varied computer pages of information at a time (see "Displays" below).
>
> (b) At least one printer for recording data both routinely and *on demand*, meaning "whenever needed".

The console may also have one or more separate digital controllers that are continuously in view for important variables. These controllers are operated in the traditional way but they also tie into the distributed control system. The central operator has access to all the information that his local operators have plus information from other areas. He is the principal operator, and he has overall control of the process operation of the plant.

The central control room may have peripheral devices for the computer, including a cartridge tape recorder

and a plotter. The room may also have a central programmable logic controller (PLC) that is connected to the distributed control system. If the logic control is only minor, the distributed control system can perform it, although use of the PLC may be preferred.

(3) *Plant Management Area.* This is an office in which plant planning and business are conducted. The area may be in an auxiliary building on the plant site. It contains a computer video screen and keyboard. The computer has access to all the information that is conveyed to or developed at the central operator's console, though generally the management does not get involved with operating minutiae.

(4) *Company Management Area.* This area is generally at the headquarters of the company, which may be located in a distant part of the country. The area contains a video screen and keyboard, and communicates with the plant through telephone lines and the distributed-control highways.

A distributed control system requires more intercommunication among its elements than do other systems. The multiplexed data highway is the key to the whole system (see Section 3-2-5). The highway may use cables of either electrical wire, which carries electrical pulses, or optical fiber, which carries pulses of light.

The highway ties together all the instruments and people who are concerned with the operation and management of the plant. It is the path through which all the multitude of electronic signals flow, then to be sorted and translated at the various destinations into suitable images, records, actions, reactions, and interactions. Relocated or additional instruments can be easily tied into the highway at any point; there is no need to run additional wires from these instruments to the control room, as would be required for the traditional centralized control. (For data highway incompatibilities, see Section 7-3-3.)

The general term *data highway* covers two subcategories: *global highway*, which is a common channel that runs throughout the plant, and *local highway*, which connects a group of devices in a local area to the global highway and may handle communications within the group. The global highway may be a simple line or a closed ring; the ring provides greater reliability because its branches remain interconnected even if the ring were to break in one place. Figure 7-5 shows a system with a single global highway and multiple local highways.

Further reliability for the system is frequently provided by fully redundant second highways for global and local service. The dual highways may both be active or one may be standing by to be switched into operation automatically if the primary highway were to fail.

Data highways may be miles in length. They may carry information for thousands of control loops, but the heavier the traffic, the slower is the system performance. Heavy traffic may require one or more additional highways.

Video displays, usually in color, are called up by keyboard. The displays for the central console include the following possibilities:

(1) *Overview.* This may be a diagram of a selected section of the process plant, showing process data. Or it may consist of graphs for process variables in a group of approximately 16 pieces of process equipment from this section of the plant, showing deviations of the variables from their set points.

(2) *Group.* This is an enlargement of a part of the overview diagram showing additional data and permitting control changes by the operator. Or it may be an expansion of information for the important process variables of one of the above pieces of process equipment, showing the changing status of the process by a graph and digital readings.

(3) *Detail.* This is a listing of tuning and alarm settings for a single selected control loop; the operator can alter these settings. Or it may be a graph of the trend of a selected variable.

The local operator's console can display these same pictures, but typically only one at a time on a single video screen.

The central console may have other features, such as a group of standby recorders that an operator can select from the keyboard to record process variables.

Some of the overall benefits of distributed control are listed below.

(1) *For the Process Operators — Rank 1.*
  (a) More informed and timely supervision of plant operations.
  (b) Quicker response in emergencies.
  (c) Automatic preparation of operating reports.

(2) *For the Plant Manager — Rank 2.* Easier and earlier reports concerning the following:
  (a) Abnormal events.
  (b) Production rates.
  (c) Materials and energy consumed.
  (d) Process efficiency.
  (e) Optimization of operations for minimum cost or maximum output.
  (f) Production costs.
  (g) Deviations from government requirements.
  (h) Historical data.

  (i) Production scheduling.
  (j) Inventories of raw materials and products.

(3) *For the Company Manager — Rank 3.* More timely information on plant production, costs, and other factors that are to be integrated into the overall business plan to maximize profit, strengthen market position, maximize cash flow, and satisfy government reporting requirements. Information can be presented in the form of three-dimensional pie charts and bar graphs.

(4) *For the Plant Laboratory.* The laboratory may have a computer connected to the distributed control data highway to receive information as follows:
  (a) Operating test data.
  (b) Chemical analyses as determined by automatic analyzers in the plant.
  (c) Other process information.
  (d) Historical data.

7-3-3. An Assessment

The coming of distributed control has caused important changes in the design and use of control systems, especially for large plants. But the resulting improvements in performance are a matter of degree, not of kind. The fanciest hardware in the world used with an inferior control strategy may bedazzle us with technology but will be outshone in process results by humdrum, old-hat hardware using the right control strategy. There is indeed a place for the latest superduper electronic wonders but there is no substitute for good control thinking, which is the bedrock of process control.

On the other hand, assuming equal control strategies for traditional hardware and for a distributed control system, the latter offers the following important benefits:

(1) Improved control of the process by the plant operator.
(2) Better communication among the company Ranks 1, 2, and 3.
(3) Greater flexibility with easy changes of control plan.
(4) Less hardware because of the electronic concentration of functions.
(5) Lower costs for installed signal lines and hardware.
(6) More reliability through extensive use of redundancy and the semi-autonomy of local instrument loops.
(7) Less control room space with a reduced need for utility services for comfort.

Traditional control systems may be made up of instruments from many different manufacturers and are frequently purchased that way. These instruments have a high degree of compatibility and interchangeability. It is up to the traditional system designer to ensure the compatibility of several instruments in a loop, for example, and usually this is not a problem.

The situation is different for distributed control, for which all the devices that are or will be *directly* linked to the data highway must come from one manufacturer and should be bought as a system in order to provide signal compatibility. Otherwise, signal converters must be interposed between the highway and other manufacturers' instruments. However, ISA Standards Committee SP50, *Signal Compatibility of Electrical Instruments*, is working to develop an industry standard for data highways and, thereby, to eliminate data highway incompatibilities.

8

External Influences On Instrumentation

8-1. EFFECTS ON MATERIALS

8-1-1. From Temperature

Instruments and their interconnecting wiring and piping, including tubing, must be able to withstand, to some acceptable degree, the effects of the process conditions that the devices may be directly exposed to. The exposure may be caused by inserting instruments into the process fluids or by piping the fluids to the instruments. This closeness may subject the devices to severe conditions — extremes of temperature, pressure, velocity, corrosiveness, and other factors related to the process fluids. All these things must be taken into account when choosing instruments and their methods of use.

Furthermore, there are environmental or ambient factors (*ambient* means *surrounding*) that must also be taken into account. These factors may be unrelated to the process or related only indirectly; for example, a pipeline may give off much heat and make the nearby surroundings hot.

All these external factors may affect the materials of the instruments or their accuracy, or both. Similarly, the factors affect the materials of wiring and piping. Some effects on instruments are discussed below.

The plant environments in which instruments may be placed range from an arctic –50°F, or lower, to a tropical 140°F, or higher (–46°C to 60°C). The minimum temperature of an instrument may be as low as the lowest outdoor air temperature. However, the maximum temperature may exceed the highest outdoor air temperature because of heat received directly from solar radiation or nearby heat sources.

ISA Standard S71.01-1985, *Environmental Conditions for Process Measurement and Control Systems: Temperature and Humidity*, [66] establishes classifications for location and severity of environmental temperature and humidity.

Protective enclosures for heating or cooling may be required or, when feasible, the instruments may be relocated.

Electrical parts, such as transistor circuits and operating coils for solenoid actuators, are sensitive to high temperature, which can shorten their effective life. The manufacturers' instructions for household electronic goods, such as a radio, often warn the buyer against keeping the goods in a hot location. Manufacturers specify different ambient temperature limits for their instruments; for example, the limits may be as follows:

(1) For operation: +40°F to +122°F (+5°C to +50°C)
(2) For storage: –40°F to +150°F (–40°C to +65°C)

For pneumatic field instruments, a manufacturer specifies an environmental temperature range of –40°F to +180°F (–40°C to 82°C).

Metals lose strength and soften as they get hotter. At some elevated temperature, their crystal structure changes, resulting in physical changes that persist after the temperature falls. The metals then require heat-treating to restore the original properties, and such treating is usually not feasible.

Metals that are chilled to very low temperatures become prone to fracture from brittleness. For example, at temperatures below –20°F (–29°C), stainless steel is often used instead of ordinary carbon steel because the stainless steel is much less subject to brittle fracture. The brittleness does not persist when the metals are warmed.

Plastic parts, such as gaskets, deteriorate faster and tend to become irreversibly brittle from chemical change at high temperatures. They become reversibly brittle from simple hardening at very low temperatures.

The passage of time causes deterioration of many materials. This aging effect occurs more rapidly at high temperature than at low temperature.

8-1-2. From Humidity

Humidity is the content of water vapor in the air and is one of the environmental use factors that instrument manufacturers specify. When liquid water evaporates, it goes into the form of vapor, which is always invisible. (A water fog or mist — loosely called "steam" — over a pond or a pot of boiling water is actually formed by minute particles of liquid water.)

The amount of water vapor that is present in air is the *absolute humidity*, expressed in pounds of vapor per cubic foot of air or equivalent units. The absolute humidity can vary from zero to some value that depends on the air temperature; the higher the temperature, the more moisture can be retained in the air. That is why the moisture in a warm room condenses on a cold outside window. At a given temperature, air is said to be *saturated* when it

contains the maximum amount of water vapor that can be retained at that temperature; it cannot dry up a water puddle.

Atmospheric air is usually not saturated, so it contains less than the maximum quantity of water vapor for the existing temperature. *Relative humidity* (RH) is the ratio of the actual amount of vapor present (per cubic foot of air) to the greater amount of vapor that would be present if the air were saturated. Relative humidity is expressed in percent and can vary from zero, which is for so-called bone-dry air, to 100 percent, which is for saturated air. *Relative humidity* is normally what is intended when a weather report speaks of "humidity". The relative humidity, not the absolute humidity, is what makes us feel comfortable or uncomfortable.

Electronic-instrument manufacturers warrant their instruments for service typically for a humidity range of 5 to 95% RH when there is no condensation. A humid atmosphere inside an instrument may cause moisture to condense on the instrument components. This may lead to instrument damage from an electrical fault (a ground or short circuit). However, an instrument containing electrical parts that are energized in normal operation receives heat from those parts; the heat lowers the relative humidity of the inside air even though it does not lower the absolute humidity. This reduces the possibility of condensation.

A humid atmosphere also promotes corrosion and deterioration of susceptible materials. Some instruments that are intended for a hot humid climate are *tropicalized*, that is, designed and treated to protect them against that environment.

8-1-3. From Explosion

Certain vapors or gases (such as benzene or methane), or dust from solids (such as coal, grain, or wood), or fibers can burn or explode if they are mixed with air and ignited. Such mixtures are called *flammable* or *combustible* although they have, illogically, also been called *inflammable.*

For a flammable vapor or gas to burn with air — explosion is an extremely rapid burning — there must be a burnable mixture. Only then can there be the chemical reaction that constitutes burning. The concentration of each flammable substance must be within a specific range in order to sustain burning. The limits of the range are only approximate because they depend on the temperature, pressure, humidity, and other components of the mixture. A few examples of the lower and upper limits of flammability are given in Table 8-1. [38]

Table 8-1. Representative Flammability Limits in Air

Flammable Substance	Flammability Limits, Volume Percent	
	Lower	Upper
Acetylene	2.5	80.0
Carbon Monoxide	12.5	74.2
Coal Dust	4.8	33.5
Natural Gas	4.8	13.5
Propane	2.2	9.5

For an explosion to begin in a mixture, all of the following must exist simultaneously:

(1) A flammable substance.
(2) Air or other oxidizing substance.
(3) Mixture must be within the flammable range.
(4) Starter source of ignition, which may be either an electric spark or a high-temperature spot.

One of the advantages of pneumatic instruments in potentially explosive areas is that the instruments do not contain an ignition source unless they contain electrical components, such as switches. Certain instruments, such as chemical analyzers, are unavoidably electrical in design. But even where avoidable, the potential hazard from explosion does not necessarily rule out the use of electrical instruments. The electrical instruments can be put inside explosion-proof enclosures or be otherwise made safe, as discussed below.

Methods of protection against possible hazards in a plant that handles flammable substances include the following.

• *Use properly designed and properly maintained process equipment, operate the equipment properly, and maintain good housekeeping.* The idea here is to avoid even the beginning of trouble.

• *Use continuous automatic analyzers.* If such an analyzer is available to detect a hazardous substance in the atmosphere, it can be used to give an early warning of the presence of the substance in case of a leak. The operator can then take protective action before the lower flammable limit is reached. Obviously, to rely on the analyzer, it must be kept in good working order.

• *Use wholly pneumatic instruments.* This eliminates the possibility of an instrument creating an electric spark in the hazardous area. Wholly pneumatic switches and alarms, both audible and visible, are commercially available.

• *Use explosion-proof housings.* (In Europe, these housings are called *flameproof.*) Such a housing for an instrument or a component does not necessarily prevent an electric spark and a resulting explosion from occurring. But if a mixture is ignited inside the housing, the explosion is confined within the housing by two means: The housing is structurally strong enough to withstand the internal explosion without being damaged, and it has a closure joint that is tight enough to permit only slow release of the resulting combustion gases. The National Electrical Code [39] has established categories for *hazardous locations*, which are defined as locations in which combustible vapor, gas, dust, or fibers may be present in explosive proportions. The classifications are:

(1) Class I, II, or III, based on the physical form of the combustible material — gas or vapor, dust, or fiber.
(2) Division 1 or 2, based on whether explosive concentrations are present or apt to be present, or whether they are contained but may be present in an abnormal situation.
(3) Groups *A* through *G*, based on the specific kind of hazardous substance that may be encountered.

Other related information is available. [40–43]

• *Prevent the explosive mixture from making contact with the ignition source.* This may be done for flammable vapors and gases in one of the following ways:

(1) The manufacturer encloses a switch or other electrical part in a hermetically sealed container.

(2) A *purge* — which is a small flow — of clean air or inert gas, such as nitrogen, is used to sweep out the hazardous substance. The purge gas pressure must be maintained high enough to prevent backflow, and adequate safeguards must be provided to guard against failure of the gas purge. Depending on the situation, other requirements may exist. [44]

• *Use intrinsically safe instrument systems.* This method is based on the fact that a combustible mixture must be heated to its ignition temperature before the mixture can explode. A spark can be so weak as to release insufficient heat to sustain combustion, so nothing really happens. On the other hand, a strong spark releases enough heat to burn enough combustibles to initiate a self-sustaining reaction, and the explosion follows. [45] To be intrinsically safe, the potential energy release must be so low as to be safe under all normal conditions and any conceivable combination of abnormal circumstances or faults.

8-1-4. *From Environment-Borne Contaminants*

There are several general types of contaminants that may affect instruments through chemical, mechanical, thermal, or electrical effects. They are as follows.

- *Airborne*

(1) *Liquids.* Liquids may be carried to instruments and their appurtenances by rain, snow, or plant washdown.
(2) *Vapors.* Vapors of water, solvents, lubricants, and other chemicals can migrate through the atmosphere. They may condense, ending up as droplets or puddles.
(3) *Aerosols.* These are suspensions of minute droplets of liquid that form a mist containing salts, as from brackish water (containing salt) or sea water. The aerosols may be created by water cooling towers, by the action of waves or wind, or by emissions from stacks.
(4) *Solids.* Dust is a universal contaminant of environmental air and other gases. The dust may originate from a large variety of sources including textile fibers, furnace fly ash, plastic powders, tobacco smoke, road dust. Some manufacturing processes, such as those for certain electronic parts, require almost perfectly dust-free air. The work is carried on in so-called *clean rooms*, which require very special design, and extreme and continuing precautions to maintain cleanliness.
(5) *Gases.* Chemical pollutants originate from transportation vehicles, industrial plants, trees and other flora, and that innocent place — the home — which releases chlorine compounds from bleach, organic cleaning compounds, combustion gases, ammonia, gaseous byproducts from burning tobacco, and ethyl alcohol from alcoholic beverages. Our source of life, the sun, does its bit through its ultraviolet rays, which convert some atmospheric oxygen to ozone, a powerful chemical reagent. The ultraviolet light also catalyzes (promotes) the interactions of chemicals and creates still other chemicals. So far as our atmosphere is concerned, we can say it is a real chemical jungle out there.

- *Biological.* Biological factors include:

(1) Birds and insects that build nests in and plug vent lines.
(2) Rodents and insects that chew away electrical insulation.
(3) Dead creatures, fungi, and mold. (Shortly after World War II, the electric computers of the day used electromechanical relays and vacuum tubes. A computer failed, and the people in charge found that the trouble was caused by the body of a dead insect that prevented the electric contacts of a relay from operating. The system had a bug in it. They removed the carcass, thus "debugging" the system, which worked again. The words *bug* and *debug*, relating to a system defect and its correction, have become universalized and are now part of the English language.)
(4) Marine life that builds deposits on sensing probes in pipes and on heat-transfer surfaces.

The pollutants, some by themselves and others by making use of the moisture in the air, can affect instrumentation and may lead to failure of part or all of an instrument system. The effects include the following types:

(1) Chemical: weakening or pitting a smooth part or creating pinholes.
(2) Mechanical: accelerating the wear of moving parts; increasing friction, thus interfering with sensing.
(3) Thermal: fouling the heat-transfer surfaces, leading to reduced heat-transfer efficiency and sometimes plugging the heat-transfer tubes.
(4) Electrical: short circuiting, grounding, or insulating electrical contacts.

ISA Standard S71.04-1985, *Environmental Conditions for Process Measurement and Control Systems: Airborne Contaminants*, establishes classifications for the type and severity of airborne contamination. [67]

To protect instruments against harmful atmospheric effects, the National Electrical Manufacturers Association (NEMA) Standards Publication No. 1S 1.1-1975 has established classifications for enclosures for electrical devices, including instruments. [78] The NEMA classifications are used for non-electrical instruments also. The classifications begin with Type 1, which is called *general purpose* and is intended for use indoors in a normally "clean" atmosphere. Classification Types 2 through 13 are for enclosures that are dust-tight, drip-proof, rain-tight, rain-proof, sleet-resistant, sleet-proof, watertight, submersible, corrosion-resistant, oil-immersed, oil-tight, or for flammable service. Enclosures satisfying these various classifications are available commercially. In some cases, a roofed open shelter may suffice.

8-1-5. *From Radioactivity*

Radioactivity is the emission of nuclear particles. It comes from radioactive substances on earth and from the sun, which is a huge nuclear furnace. It exists virtually everywhere, including the home, as weak background radiation. In normal everyday doses, it may be harmful to people and things over the long term but the extent of the harm, if any, is under scientific debate.

Radioactivity can become a problem in a plant that generates nuclear power or that processes nuclear substances for fuel, medical purposes, or weapons. Small quantities of nuclear substances are also used occasionally in ordinary, non-nuclear process plants to measure the density of materials or the level in a vessel, also to check the quality of metal welds. In such uses, the local intensity of nuclear radiation may be high and damaging; there is no scientific argument about this. Radiation warning signs should be placed where this hazard exists.

The effect of radioactivity on organic materials — plastics, oils, and others — depending on the material, is to reduce strength, increase hardness and brittleness, cause swelling, discolor, make gummy, reduce lubricity, and induce radioactivity in the materials. Metals may suffer embrittlement, change of strength but not necessarily a reduction in strength, transmutation so that a homogeneous metal ends up as an alloy, and the metals may become radioactive.

Instruments that will be used in a nuclear environment should use only materials, so far as possible, that can withstand the expected radiation over an acceptable period of time. For example, steel or stainless steel is usually acceptable, while aluminum, mercury, and zinc are not acceptable. Polypropylene plastic is acceptable, Teflon™ is not acceptable.

If a suitable long-life material is not available for an instrument part, an appropriate replacement schedule should be followed to ensure proper functioning of the part.

8-1-6. *From Shaking*

Instruments are mounted on supports: pedestals, boards, piping, walls, machinery, and so on. Machines and motors vibrate and cause the structural members of the building to vibrate; this causes everything attached to the building to vibrate. A pipe may vibrate from fluid flow or from an unstable control loop whose control valve *hunts*, that is, oscillates; this causes everything attached to the pipe to vibrate. Whatever the source of vibration, the instruments often vibrate to some degree.

Plant vibration mechanically wears the moving parts of instruments. It also causes *fatigue*, which results from repeated or prolonged stress and which may cause a part to crack or break. A familiar example of failure from fatigue is the breaking of a paper clip by bending it repeatedly. In most cases, instrument vibration is small enough so that fatigue failure does not occur.

When an earthquake occurs, it creates a shaking that differs from the normal plant vibration. The swings are much wider and usually slower. This shaking does not cause fatigue inasmuch as it does not last very long, but it may be so severe as to cause permanent deformation or breakage of parts of the instruments or their supports.

To avoid instrument damage from the shakings of normal vibrations and of earthquakes, many instruments are available to withstand severe movements and shock without failing. Shock absorbers are sometimes used.

Instrument tubing, electrical cables, and their supporting trays and base supports must also withstand plant shakings.

The ability of instrument systems to keep functioning during and after earthquakes is very important for nuclear power plant systems that are related to nuclear safety. The designers of these plants and manufacturers of the associated instruments take great pains to provide overall system integrity. The same kind of care may be appropriate for important systems in other kinds of plants.

8-1-7. *From Lightning*

Outdoor instruments may require protection against being struck and damaged by lightning. All electrical instruments, especially those outdoors, may require protection to prevent internal damage from voltage surges caused by lightning.

8-2. EFFECTS ON ACCURACY

8-2-1. *From Temperature*

Ambient temperature may affect the properties of mechanical parts, such as a bellows, or electronic parts. This may result in a loss of instrument accuracy as the temperature changes from its design value. Two typical manufacturers' statements of the effect that ambient temperature changes have on the accuracy of their instruments are as follows:

- ±0.28% of reading per 50°F (28°C) change, based on a reference temperature of 77°F (25°C)
- ±0.25% of span per 50°F (28°C) change

Pressure or differential-pressure measurements are frequently made with their sensing lines filled with liquid to measure flow or level. These lines are subject to ambient temperature changes from day to night and from changes in the weather. The changes heat or cool the liquid in the lines, thereby changing the liquid density and the resulting pressure that the instruments sense. Even if the instrument itself is protected and kept at a constant temperature, the effect on the external sensing line causes measuring errors that may be significant unless precautions are taken.

Changes in ambient temperature raise or lower the temperature of the capillary and pressure sensor in a filled system, which may be used to measure process temperature, pressure, or level. Since a filled system is really a pressure system (as explained in "Filled Thermal System," Section 3-7-2), the result is to change the measurement. Ambient temperature effects can be nullified if the instrument is furnished with automatic temperature compensation.

All the temperature-induced errors are reversible and disappear when the ambient temperature returns to its design value.

8-2-2. From Humidity

Ambient humidity (see Section 8-1-2) may cause electrical leaks that do not necessarily stop electrical instruments from functioning but that are hidden causes of instrument errors.

8-2-3. From Pressure

Ambient pressure affects the accuracy of gage pressure instruments (see Section 3-4-2).

8-2-4. From Environment-Borne Contaminants

These contaminants were discussed in Section 8-1-4. Ways that they may affect instrument accuracy include causing leakage of electrical signals, increasing mechanical friction, interfering with electrical switches, plugging vents, and slowing the dynamic response of temperature sensors because of the buildup of external films.

8-2-5. From Lightning and Other Electrical Phenomena

Lightning may cause transient voltage surges or momentary loss of voltage in electrical circuits. These effects may cause a loss of circuit memory or a change of information in electronic storage. Protection against the effects may include the use of surge suppressors and backup power supplies.

Electronic systems may also be affected by nearby electrical noise — so-called *radio frequency interference* (rfi) and *electromagnetic interference* (emi) — which may be produced by electrical power cables, power line fluctuations as machines swing on and off, radiotelephones, and portable tools. Protection against these effects may be provided by noise filters and proper electrical wiring practices, including shielding, grounding, and barriers against or separation from noise sources.

8-2-6. From Gravity

Certain measurements — of pressure and other variables — may conceivably be in error by as much as 0.5 percent of the true value because of gravity differences. This is rare but could occur if the sensing instrument is used in a geographic location that is very far from the place where it was calibrated. When calibrating an instrument by means of a standard weight, correction should be made for the effect of gravity on the weight, if appropriate. (See the discussion of *force* and *weight* in Section 3-4.)

8-3. POWER-SUPPLY REQUIREMENTS

Power supplies — electrical, pneumatic, or hydraulic — are needed to enable most instruments to function. For the instruments to function properly, the power supplies have to meet the manufacturer's specifications, as follows.

8-3-1. Electrical Power Supplies

Representative requirements for voltage and frequency are:

(1) 120 V, +10% to −15%, ac and 50/60 ± 2 Hz (*hertz*, formerly *cycles per second*)
(2) 125 V, +15 to −20 V, dc
(3) 24 ± 2 V, dc

8-3-2. Pneumatic Power Supplies

A representative manufacturer's specification for a pneumatic instrument calls for the air supply pressure to be:

(1) Normal: 5 psi above the maximum pressure of the instrument output.
(2) Minimum: 3 psi above the maximum pressure of the instrument output.
(3) Maximum: 50 psig.

The purity of the instrument air may be important, especially if an instrument has small air passages. The air supply comes from a compressor whose air discharge is usually relatively humid and often carries droplets of compressor oil and dust particles; this air is called *plant air*. The droplets and dust may coalesce or agglomerate; the condition is worsened when condensation or freezing occurs in the air system. The result may be to clog the air passages of pneumatic instruments so that the instruments operate poorly or not at all.

The normal practice for pneumatic instrumentation is to provide an air-purification system. This is required even in most electronically controlled plants, which usually require pneumatic control valves and some local pneumatic instruments. The product of the purification system is dry and clean *instrument air*. For instrument air, the word *dry* means to have a relative humidity so low as to prevent condensation of the moisture in the air at the lowest plant temperature to which the air may be exposed.

The purification system is supplied with plant air, which is passed through filters to mechanically remove liquid and solid particles, and driers to remove water and other vapors. Some driers dry by using either refrigeration or expendable chemical agents, such as lime. Another type of drier uses desiccant (drying) pellets of alumina or silica gel that retain the vapors from the air passing through the drier; this type of desiccant can be regenerated by heat and used repeatedly. (We sometimes encounter small bags of desiccant when we open a new package containing an expensive camera or pills that have to be kept dry.)

ANSI/ISA Standard S7.3-1975 (R 1981), *Quality Standard for Instrument Air*, establishes allowable limits for moisture, particle size, oil concentration, and other contaminants. [69] The American Nuclear Society (ANS) Standard 59.3-1977, *Safety Related Control Air Systems*, discusses instrument-air systems for use in nuclear power plants. [70] However, this standard may be helpful also for non-nuclear plants, especially for those process systems having a particular need for reliability of their pneumatic instrument systems.

There are situations where air substitutes may be used. An instrument far from a plant air supply may use nitrogen, which is a chemically inert gas, supplied in portable steel bottles with pressure regulators. Even pipelines carrying process gases that are dry and noncorrosive are sometimes used to power local pneumatic instruments.

Plant air is sometimes used for pneumatic devices such as on-off piston actuators; these actuators are simple machines that may not have small passages that can become clogged. But plant air is not customarily used for such applications unless the point of use is remote, instrument air is not available, and plant air is available.

8-3-3. Hydraulic Power Supplies

A hydraulic instrument system uses a hydraulic liquid that has to be kept clean and at the operating pressure specified by the supplier of the hydraulic system.

9

Incidental Hardware

The previous chapters described the hardware scenario in which instrumentation systems function. The instruments are the lead actors in this production, whose purpose is to show how a process plant can be controlled. The roles of supporting cast and bit actors are played by other hardware that is important in making the production a hit and bringing the audience to its feet, cheering. Prominent among the many secondary members of the cast are the following devices.

9-1. TEMPERATURE WELLS

Temperature wells, commonly called *thermowells* (in Great Britain, *pockets*) are metal pieces, as shown in Figure 9-1. They are customarily supplied as accessories for insertion-type temperature sensors. The well is installed projecting into a process pipeline or vessel, and the sensor is inserted into the well. The well is made from a corrosion-resistant material and is strong enough to resist mechanical damage and to stand up under the process pressure and temperature. The purposes of the well are as follows:

(1) To protect the temperature sensor against corrosion, breakage from vibration caused by fluid flow, and damage from erosion.

(2) To enable the sensor to be replaced without requiring the plant operation to be stopped.
(3) To permit future insertion of a sensor for either temporary test measurements or permanent use.

Figure 9-1. A Typical Thermowell

Thermowells have thermal resistance, which is undesirable because it slows the response of the sensor to temperature changes.

9-2. ISOLATION VALVES

With few exceptions, every sensing line is provided with a *root valve*, which is an isolation, or shutoff, valve close to the process line or vessel. Generally, each sensing line branch running to a separate instrument has a shutoff valve placed near the instrument; this valve is known as an *instrument valve*. Each sensing line branch may also have a valve to bleed air or other gas from the branch if it carries a liquid, or to bleed liquid if it carries a gas; the purpose is to keep each line or branch full of the intended fluid. A valve may be provided for connecting a calibration or test instrument. Figure 9-2 shows a typical sensing line for a pressure instrument.

Similarly, there are shutoff valves for pneumatic or hydraulic signal lines and power supplies.

Analogously, there may be isolation switches for electrical signal lines, and circuit breakers or switches for power supplies. Electrical instruments may have connections such as screws or jacks (a form of electrical socket) for temporary use with calibration or test instruments.

Figure 9-2. A Typical Pressure Instrument Installation

9-3. WEATHERIZING DEVICES

Many process liquids freeze or precipitate solids in cold weather; for example, sodium chloride (common salt) begins to settle out from a 36.6-percent solution with water unless the temperature is kept above 122°F (50°C). Some liquids boil in hot weather. These effects of extreme temperatures may present a problem for sensing lines, especially if they are stagnant, or for the instruments themselves. To cover these devices with thermal insulation is not an answer because insulation does not supply or remove heat; it only slows the cooling of hot liquids or the warming of cold liquids.

It is common practice to provide a heating system to protect susceptible individual lines and instruments against the outdoor cold. Such protection is called *winterizing*. Methods for doing this are:

(1) *Electrical Tracing.* Electrical heating elements are run alongside and in close contact with the sensing lines and instruments.

(2) *Electrical Heating.* Metallic sensing lines themselves are made to act as heating elements by passing electricity through their walls.

(3) *Steam Tracing.* Small tubing is laid in contact with the lines and instruments to be heated, and the tubing carries heating steam.

(4) *Enclosures.* Special enclosures are very often used for the instruments. The enclosures are heated by either electricity or steam.

Each kind of protective system needs some auxiliary instrumentation for regulating the heat input to the process instrumentation that is to be protected.

Some lines and instruments may contain a volatile liquid that boils on a hot day. Such cases require a means for cooling — *summerizing*. The simplest practice is to change to a sensing technique that does not require the sensing lines to contain the volatile liquid.

9-4. INSTRUMENT ENCLOSURES

Section 8-1-4 discussed a NEMA Standard, No. 1S 1.1, for enclosures to protect all types of instruments against harmful atmospheric effects.

9-5. POWER SUPPLIES

Power supplies for instruments are usually provided by engineering disciplines other than control systems engineering or instrument engineering. This assumes that the operating company or the consulting company doing the engineering for the plant is large enough to have separate departments or groups for the various disciplines. Small organizations may instead have one or more jacks-of-all-trades but they tend to lack the expertise gained from full-time specialization.

The amount of electrical power required for plant instruments and for electrical tracing is typically based on a count of the various types of instruments and the amount of tracing, as determined by the instrument department. The electrical department uses this information to plan their electrical systems and to procure the necessary power equipment, cables, and other specialized hardware.

Similarly, the instrument department counts the number and types of pneumatic instruments, including control valves, and determines the amount of air that is required. The mechanical department then procures the requisite air compressors and equipment to dry and clean the air. The design of the air distribution system is typically a complementary effort by the piping department, for major air piping, and by the instrument department or construction department, for the local air piping and tubing.

The instrument department also estimates the amount of steam tracing that will be needed, and the mechanical department allows for this in designing the plant steam system.

Instrument air is usually distributed at a pressure of nominally 100 psig, which is lowered at each instrument as required, for example, to 20 or 35 psig. The pressure reduction is usually performed by a 1/4-inch adjustable reducing valve having a filter, drain connection, outlet pressure gage, and overpressure protection. This valve is termed an *air set* or *air filter-regulator* and is procured by the instrument department, either as an accessory for each instrument or as a bulk commodity.

If a hydraulic control system is required that is not part of a machinery package, it may be procured by the instrument department.

9-6. INSTRUMENT SUPPORTS

Supports for instruments obviously have to be strong enough to hold the dead weight of the instruments and their immediate accessories, including a portion of adjacent connecting piping, tubing, and wiring. The support design may also need to allow for the weight of careless construction or maintenance workers who step on the supports.

In addition, the effects of dynamic loads on the instrument supports should be considered, as discussed in Section 8-1-6. The supports are usually designed by civil engineers.

9-7. INSTRUMENT BOARDS

Instrument boards come in different sizes and styles, which are summarized as follows:

(1) *Consoles,* which were discussed in Section 7-3.
(2) *Instrument racks,* which are open steel frameworks on which instrumentation is mounted.
(3) *Cubicles,* which are large enclosures.
(4) *Panels.* The term is frequently used ambiguously. ANSI/ISA Standard S5.1 (R1981) states that *panel* and *board* are synonymous, and "panel" is often used in the sense of "board". However, both "boards" and "panels" have component sections that are called "panels". For clarity, this book refers to an entire structure as a *board* and its sections as *panels*.
(5) *Benchboards,* which are vertical boards that have a sloping, desk-height section.

10

Design and Selection of Instruments and Systems: Engineering Considerations

We have seen how process instruments are used, singly and in combination. But before deciding on a particular kind of instrument or loop for measuring or controlling, there are engineering factors, described below, that should be taken into consideration. There are also managerial factors, which will be discussed in Chapter 11.

10-1. LOOP ACCURACY

In Section 3-1-5, we took up the subject of accuracy of a single instrument. (We should remember that when the word *accuracy* is used for an instrument or loop, its meaning is *error*, *inaccuracy*, or *uncertainty*. For clarity, the word *error* is used instead of *accuracy* in the following discus-

sion.) What is the overall error of a measurement that is made through a series of instruments — for example, the instruments LE-1, LT-1, LY-1C, and LC-1, as shown in Figure 3-8?

If the loop exists physically, the way to determine its actual error is by testing the loop. This is done by giving the first element of the loop an input and seeing what comes out of the last element.

However, it may be necessary to predict a loop error. Suppose the errors in this measuring loop are as follows:

Sensor	LE-1	±0.5 % of span
Transmitter	LT-1	±0.25% of span
Signal converter	LY-1C	±0.75% of span
Receiving element of LC-1		±0.4 % of span

For simplicity, the above-listed errors are those claimed by the instrument manufacturers. If the environmental conditions are not as specified by a manufacturer, then additional errors are introduced, as discussed in Section 8-2, and the result becomes worse.

The overall error of the loop can be determined by one of the following methods:

• *Algebraic Error.* The simplest way to determine the overall measurement error is by adding the absolute value of the four individual errors, which means disregarding the ± symbols for the addition. The result is the *algebraic error*, which in the above example is ±1.9% of span.

This method is fast and easy. It is also very safe because it contains two implicit assumptions about the four errors:

Assumption (1): The errors are all positive or all negative, not mixed.

Assumption (2): Each error is at an extreme of the manufacturer's guaranteed error band.

The law of chance for random errors does not usually work this way, and it is unlikely that the worst-case assumptions will hold true in a given case in real life.

If an overall error of ±1.9% of span is acceptable for the service, then the matter can rest. However, if the service requires that the overall error not exceed ±1.1% of span, then the result is not acceptable; a more accurate error estimate is required.

One alternative, if the required measurement range covers only a portion of the range for one or more of the instruments, is to calibrate the instruments over just that portion instead of over the full instrument range. This reduces the overall error.

Some environmental errors may be reduced by relocating the problem instrument or instruments or by using some type of compensation.

• *Probable Error.* The most widely used method for determining the overall error gives the *probable error*, not the extreme and unlikely algebraic error. The *probable error* is the error that is just as likely to be exceeded as not to be exceeded. This is calculated by taking the square root of the sum of the squares of the individual errors, which gives a result of ±1.0% of span. This is acceptable if the requirement is for the error to be not greater than the allowable maximum, ±1.1% of span.

In this example, all the individual errors were given equal weight, that is, all were treated the same way in the calculation. In calculations of probable error for some instrument loops — frequently for flow measurement — some of the individual errors are given more weight or less weight than the other errors receive. The need for doing this depends on the mathematics of the measurement equation.

• *Statistical Error and Probabilistic Error.* In a given case, the probable error may be unacceptably large and the algebraic error is even larger. Or the very concept of probable error may be unacceptable because, by definition, the concept acknowledges a one-out-of-two chance that the actual error will be worse than the probable error. If the measuring system cannot be improved to reduce the error, then more refined methods may be used to determine the overall error.

Such methods are the *statistical method* and the *probabilistic method*, which are quite sophisticated and may involve special testing and fallible human judgment. These methods are used rarely and only when dealing with especially important cases. They yield what may be the most realistic estimates of error, but these will not necessarily be any more acceptable than the others were.

10-2. RELIABILITY

10-2-1. What Is Reliability?

Reliability is the probability that a product will provide a *specified performance* when operating under *specified operating conditions* for a *specified length of time*. In other words, the reliability of an instrument or instrument system is the consistency with which it measures or controls as it is supposed to under the right conditions and according to its program and adjustments.

Two important terms concerning reliability are the following:

(1) *MTBF*, which is the *Mean Time Between Failures* of a given type of instrument or system, as determined by test, experience, or both. "Mean time" means the "average time". A large MTBF is good and depends on (a) the equipment manufacturer using high-quality materials, conservative design, and care in manufacture and (b) the operating plant applying the equipment to the kind of service for which it was designed, and performing routine maintenance as recommended.

(2) *MTTR*, which is the *Mean Time To Repair* a failure, as determined by experience. A small MTTR is good and depends on (a) the manufacturer designing the equipment for ease of repair and (b) the operating plant having on hand or on call an adequate source of spare parts and repair people with the required skills, and easy access to the equipment needing repair.

By using MTBF and MTTR, we can determine the *availability* or *time-availability*, *A*, of an instrument or instrument system, which is the fraction of time that the instrument or system can be expected to be ready for use and to perform properly. *A* equals MTBF divided by the sum of MTBF and MTTR. The availability increases if MTBF increases or MTTR decreases.

Quite often, a manufacturer cannot provide data for MTBF and MTTR for calculating the availability, particularly for non-electrical devices. However, it can be worthwhile to attempt an estimate of the availability or to judge the apparent quality of the devices. By taking reliability into account when choosing an instrument or planning a system, the plant will tend to have lower annual repair costs for faulty instruments. Also, plant downtime because of instrument failure and the resulting losses from non-production will be minimized. These factors may be influential in justifying a purchasing decision in favor of equipment that will have the lowest cost over the life of the plant even if its initial cost is not the lowest.

10-2-2. Number of Loop Elements

Reliability is improved by reducing the number of links in the instrument chain. Section 10-1 discussed the overall accuracy of a specific measurement loop consisting of four instruments — sensor, transmitter, signal converter, and receiver. Each instrument has some risk of failure and contributes to the risk of loop failure. The loop risk can be reduced if we arrange the transmitter output signal to match the requirement for the receiver input; then we can eliminate the signal converter and its risk of failure. For this particular level loop, eliminating the converter would also lower the overall loop error from ±1.9% of span to ±1.15%, calculated by the algebraic method.

The principle of direct communications, or the lack thereof, and its effect on reliability are further exemplified in the children's game of *Telephone*, which is played as follows: A group of children sit in a line. The first child whispers a message into the ear of his neighbor, and the message is thus relayed down the line. The point is to see how distorted the message has become when it gets to the last child. The more children there are in the line, the greater is the distortion and the greater the laughter.

10-2-3. Redundancy

If an instrumentation failure in a plant could result in an unacceptable risk of physical danger or monetary loss, then redundancy may be called for. *Redundancy* means providing an alternate element to take over if the first element fails. Redundancy may be applied to any kind of equipment: sensors, controllers, computers, power supplies, fans, heat exchangers, entire systems, piping, etc. A specific example of redundancy is the dual-channel data highway used in a distributed control system, as described in Section 7-3-2; if one channel fails, the other carries on while the failed channel is repaired.

For redundancy to be fully effective, each channel should operate totally independently of the other. This means that no single misoperation, such as incorrectly opening or closing one switch, and no single failure, such as a failure of a common power supply, could disable two channels. The failure of a common power supply is an example of a *common-mode failure*. But common-mode failure may also be caused if a single falling object disables two channels by damaging two redundant controllers, for example. To prevent such failure, the two channels should be protected by separation: spacing the channels far apart from each other or providing adequate barriers betweem them.

Another way to increase plant reliability is by diversity. This calls for redundant channels to do the same thing but in different ways. The different channels are then unlikely to suffer all of the same kinds of failure. For example, a level measurement may be made by a displacement-type instrument and a head-type instrument, which measure level by different physical principles and are constructed differently.

A good design principle to follow in all plants is to separate the normal control function from a safety function. Figure 10-1(a) shows how normal control of tank level and the safety shutoff of a pump to prevent flooding can be accomplished. The level controller modulates the flow so that the tank does not normally flood. In case of high level from a severe upset, the high-level switch stops the pump and the tank inflow if the level becomes excessively high. The tank does not flood.

All parts of the scheme will probably work as they are supposed to whenever they are needed. However, the scheme has a weakness that can potentially cause trouble: Both the controller and the level switch depend on the single level transmitter and are, therefore, subject to common-mode failure.

Figure 10-1(b) shows a more reliable way to protect against flooding. Everything is the same as before except for the level switch, which now senses the tank level directly. The control loop may fail but the switch is not affected. On the other hand, if the switch fails, the control loop is not affected.

An example of good reliability is that of a man holding up his pants by using a belt and suspenders. If one device fails, he expects the other to keep the pants from falling down. His safety system has redundancy, diversity, and separation.

Some processes have problems in keeping sensors in good working order. The sensors tend to foul, for example, and become unreliable; assume they give a false high reading. Redundancy may be applied by using three separate channels to actuate an alarm if any channel reading is high. The operator can then investigate and use his judgment about shutting down the process. But, if any two channels read high, there is an automatic shutdown. The alarm is initiated by what is known as a *one-out-of-three voting system* and the automatic shutdown by a *two-out-of-three system*. The idea is that a single high reading could be an aberration and false and should not necessarily call for drastic protective action. But if that reading is confirmed by a second reading, then the high readings are considered valid. An even more conservative scheme uses a two-out-of-four system, which has four channels instead of three.

Another system measures disparities between two or more process instruments that should be giving the same reading. If a disparity becomes excessive, it actuates a disparity alarm even though a process trouble may not be apparent.

Measures to increase reliability may be applied to any process system of great potential hazard, although they have their heaviest use in the nuclear power industry. [48, 49]

There is a movement in the world of electronics, including instruments, toward the development of so-called *fault-tolerant equipment*, which has internally redundant components or circuits. The effect is to enable the instrument or system involved to continue functioning properly even if some instrument part were to fail. This technique is used extensively in some distributed control systems.

For important process systems, a *failure analysis* may be made. This is a detailed study of what may happen to the process if the various pieces of process equipment and instrument systems undergo all credible combinations of failure and mishap. The study may reveal a need for backup equipment, a change to a different type of equipment, a change of fail-safe action (see Section 5-1-4), or other changes; or it may simply confirm the adequacy of the existing system design.

10-2-4. Trustworthiness

Instruments do not get bored, their attention does not wander, they do not listen to a ballgame, they do not stop to eat or go to the rest room, they do not come to work tired, they do not have emotional problems, they do not abuse their bodies or their minds. They are always present and always attentive.

A process operation may be carefully and skillfully engineered to operate most efficiently or most safely in one particular manner. Operations people, sometimes at a low level, have been known to decide, without understanding the details and subtleties of the original design engineering, that the process should operate differently, and they change the control system accordingly. The process is thereby degraded. However, instrument systems can be arranged to have varying degrees of security to guard against tampering.

In short, barring malfunctions or tampering, for which various levels of protection can be provided, a properly designed instrument system can be counted on to do its job exactly the same way every single time. It is trustworthy in a way that no human being can match, and it contributes importantly to the reliability of the plant.

10-3. HUMAN FACTORS

In days gone by, instrument boards were merely an assemblage of panels on which instruments were placed,

(a) Less reliable

(b) More reliable

Figure 10-1. Preventing a Tank from Flooding

generally with some reasonable degree of logic and common sense. This was done according to the individual case and without a systematic approach to improving the performance of the man/machine combination. Then followed centralized control, with its mushrooming amount of information for an operator to supervise. Later came nuclear power plants with their publicized special emphasis on correct operation.

The process industries became more involved in a discipline known by various names, including *human factors engineering, human engineering, ergonomics,* and *biotechnology.* The industries were able to draw on principles developed by the military, who use large quantities of technical equipment and manpower. When the military mean business, their stakes are very high, so they have been active since World War II in developing the principles and practices for humans to make more effective use of themselves and their machines of defense and offense.

Human factors engineering has been applied to things as simple as a screwdriver, resulting in a screwdriver handle shaped as a ball instead of the traditional plain or fluted cylinder. The ball enables the person using the screwdriver to grip it more strongly, provide more twist for turning the screw, and be more comfortable while doing this. However, even with the simple new design, there are subtleties: What is the optimum size of the ball handle for most people? What surface finish shall be used for maximum grip with minimum discomfort?

The work of a control room operator involves many factors, including the following.

● *Use of the Human Body.* This is covered by *anthropometry,* which is the study of the measurements of the human body, its ability to see and hear, its tendency to fatigue, etc. This information is based on extensive research and testing.

These body characteristics plus the nature, frequency, and difficulty of the physical and mental tasks that the operator must perform, the position of his body, and his mobility requirements are important in designing instrument boards. They affect the dimensions and instrument layout of the board, even the appearance of the instruments.

Little things can affect an operator's performance over an eight-hour shift. For example, for a video screen: How good is the picture resolution, how high is the screen placed, is it aimed properly, is it protected against glare?

Other anthropometric factors include the following:

(1) *Physical Environment.* This includes items like room temperature, humidity, lighting, noise, and the amount of traffic in the room.

(2) *Instrument Arrangement.* Principles for arranging instruments on a board for the operator's benefit include the following:

(a) Instruments should be grouped logically to show their relationships to the process.
(b) Controllers or control stations and their associated readout instruments should be next to each other, so far as practical.
(c) Similar groups of instruments should be arranged similarly whether left and right or up and down, as illustrated in Figure 10-2.
(d) No more than five similar-appearing instruments should be in a line with uniform spacing unless there is color coding or other striking differentiation between neighboring instrument groups. The coding may be of panels or instrument bezels, for example. The *bezel* is the rim that holds the glass or plastic covering on the face of an instrument.

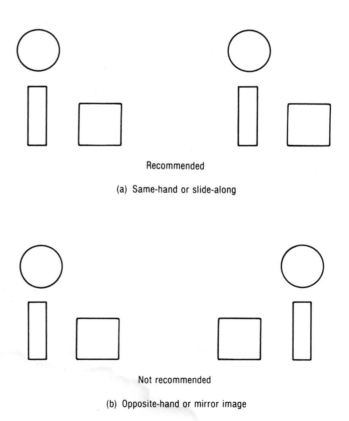

Recommended

(a) Same-hand or slide-along

Not recommended

(b) Opposite-hand or mirror image

Figure 10-2. Arrangement of Similar Instrument Groups

• *Presentation of Information.* In a large and complex plant, providing information by exception is useful to make life easier for an operator and better for the plant. Information that is routinely presented to the operator should be limited to what he needs for his routine tasks; more than that is a distraction. If trouble arises, then a second level of information should become available for the trouble area. The hierarchy of video displays — overview, group, and detail — that can be provided in a distributed control system is helpful in permitting an operator to zoom in on what is important at any moment (see Section 7-3-2).

Red is our traditional color to signal trouble. We have red traffic lights and red signal flags that mean STOP or DANGER. Red ink on the bottom line of a profit-and-loss statement for a business is bad news: It dramatically tells us that the business is losing money and is said to be "running in the red".

By the same token, an instrument board should never show any red light when everything works as it should and conditions are normal. For example, the color red should not be used for a pilot light to signal POWER ON as a normal condition; a different color should be used. If a red light ever does show up, it should have only one possible meaning — TROUBLE — and it should require a very prompt corrective response by the operator. Especially in an emergency, he should not have to lose the time to sort out whether the new red light is one of the abnormal red lights on the board or one of the normal red lights, which he will very likely find to be of no significance at this time.

Another illustration of human factors engineering in presenting information is the use of different alarm sounds for different process groups — for example, using a horn for one group of systems, a buzzer for a second group, and chimes for a third group.

There are ISA, military, and other recommendations that provide more detailed information on the above and other points. [50,51]

10-4. STANDARDIZATION OF HARDWARE

When specifying instruments for purchase, it is worthwhile to consider to what extent the choice of kinds of instruments and their suppliers should be limited in order to standardize the hardware within the plant. In this context, the word *standardize* can mean the following things:

(1) Where it is technically reasonable, handle similar situations in a plant in the same way.

Where appropriate, measure various flow rates, for example, with the same kind of instruments. However, costs should be taken into account so that the kind of flow instrument used for small pipelines may end up different from that for large pipelines. Different fluids and different accuracy requirements may reasonably justify or mandate deviations from the standard.

(2) So far as practical, specify only a certain brand or brands of instruments. Keeping to a prescribed brand for a given plant may offer the following advantages:

(a) The spare-parts inventory can be smaller because of the commonality of parts among instruments. This results in savings of investment in inventory and storage space.

(b) Many instruments are interchangeable so that a sudden need for a replacement is more apt to be filled by an existing instrument.

(c) The instrument installation people and service people work with already familiar devices. They have less need to refer to installation and service manuals, and they may work quicker and better.

The primary consideration must always remain to procure instruments that the applications truly require, not to adhere mindlessly to arbitrary purchasing rules.

10-5. INSTRUMENT SPECIFICATIONS

The planning and design of instrument systems at some stage require purchasing, or other means of procuring, the instruments and auxiliary hardware that are needed to carry out the designs. When we go to buy something, we need to tell the seller what we want him to sell. The description is normally stated on a document called a *specification, specification sheet, data sheet, requisition, bill of materials,* or otherwise. The document may refer to auxiliary documents — drawings, lists, codes, or other; if it does, the repetition of details on different documents should be avoided so far as practical. Repetition opens the door to errors and inconsistencies.

There may be sound technical or practical reasons for requesting certain non-standard features on instruments. For example, the plant maintenance department may have found that a certain type of lubrication fitting for control valves can save the plant money over the years by facilitating the servicing of the valves. Therefore, a requirement for the valve manufacturer to have these fittings installed at the factory could be justified.

Nevertheless, it is common to find overspecification of instruments by the purchaser, that is, demanding more physical or functional features and details than are specifically needed. This can apply, for example, to an electrical connection for a power supply. Certainly the instrument needs a connection, but must the purchaser demand that the connection have a 3/4-inch thread? The manufacturer's standard connection may be a 1/2-inch thread or some other conventional type. He may be able to furnish what the customer requests but the order may then become special. There are many other ways of overspecifying, and all of them tend to cause one or more of the following undesirable and avoidable results:

(1) Increased cost.
(2) Lengthened delivery time.
(3) More mistakes by the manufacturer.
(4) Shutting out a good potential bidder who cannot or will not deviate from his standard design.

There are rules that, when intelligently applied, help to reduce the overspecifying:

(1) Specify every detail that is needed and that may cause trouble if it is not specified. For example: To specify simply that an instrument should be enclosed is not sufficient if a NEMA type 4 enclosure (watertight and dustproof) is required.
(2) Do not specify anything that is not needed. Do not go into more detail than is necessary.

In short, the first thing to do is think. Then be willing to accept the manufacturer's standard product unless there is a specific reason not to.

11

Managerial Considerations

Suppose we are in charge of a production plant and we believe that the plant will operate better if it has a better process control system with new and expensive chemical analyzers and an elaborate sampling system. We are now considering whether or not we should make the change. We are not in business for our health, so we will use the old standby — money — as the criterion for our decision. We will not base the decision on vanity factors, such as "being on the cutting edge of progress". The hard question is: Will we gain more by the new system than we will lose?

Both benefits and costs of the proposed system are affected by the anticipated economic life of this system as compared to that of the existing system. Also to be considered are the effects of tax regulations, present and future, and other government policies relating to monetary inflation, interest rates, and even international trade. Governmental actions are imponderables, being based on human fallibility and passions and the self-appointed right of government to manipulate the economic affairs of people and companies. These actions create a high degree of unpredictability in investment planning. However, the advantages of some considered investments are clear-cut enough for making a decision.

11-1. BENEFITS OF THE PROPOSED SYSTEM

Regarding benefits of the new system, the following questions arise:

(1) *Greater Stability.* Will it provide better control of the plant operation and reduce the number and magnitude of harmful process upsets?

(2) *Improved Quality.* Will the system improve product quality, thereby reducing or eliminating the cost of reruns or wastage? Will it provide a higher-grade product for which a premium price may be charged?

(3) *Improved Production.* Will it increase production and sales?

(4) *Higher Efficiency.* Will it raise plant efficiency, resulting in higher production or lower requirements for feed materials and energy?

(5) *Improved Reliability.* Will it be more reliable in keeping the plant operating and avoiding the costs of downtime and repairs?

(6) *Greater Safety*. Will it be safer and cleaner? Will it help to avoid penalties for violation of governmental regulations? Will the system lower insurance costs?

11-2. COSTS OF THE PROPOSED SYSTEM

The other side of the evaluation concerns the costs for the new system, with the following questions:

(1) *Cost to Buy and Install*. How much will the system cost to buy and install?

(2) *Plant Downtime*. Will the system require plant downtime for installation? What will this cost?

(3) *Manufacturer's Field Service*. Will the system initially require field service by the manufacturer? How much will this cost?

(4) *Future Service*. What will be the increase for the future cost of the servicing, repair, and the associated plant downtimes for the new system as compared to the future cost, if any, for the present system?

(5) *Incidental Plant Changes*. What incidental plant changes will be required for the instrument system, and what will they cost? The changes may include added piping, expansion of the power supply system, structural modifications, additional or altered instrument supports and boards, etc.

(6) *Future Operating Cost*. Will there be a significant change in operating cost to use the new system?

(7) *Availability of Investment Capital*. Where will the money come from to pay for the new system? What will the money cost?

An additional factor that may have bearing on the costs has to do with who shall do the work of instrument system design, purchasing, and installation, which raises the following questions:

(1) *Design*. Who shall do the detailed design for the contemplated instrument system? If a new process system design is purchased, shall the process design company design the instrument system? Or shall the instrument system be designed by the headquarters staff of the operating company? An engineering contractor? A full-line instrument manufacturer?

(2) *Purchasing*. Shall the designer of the instrument system supply the system? If the instrument system is both designed and purchased by the company that designed the system, there is a single source that is responsible for the process and instrument package. If an operating problem arises, that company supposedly cannot point a finger at someone else as being responsible.

Shall our operating company purchase the required instrumentation? The responsibility for purchasing may carry with it the responsibility for coordinating with the company that designs the instrument system and for performing inspection work that may be necessary. Does our company have the manpower and skill to handle this additional work?

(3) *Installation*. Shall the instrumentation be installed by our operating company? By the instrument supplier? By outside contractors, with construction management performed by the operating company? By a general contractor who will subcontract and manage as required?

(4) *Labor Relations*. Are we compelled to use union labor?

(5) *Coordination*. What is the most effective arrangement overall for coordinating the different phases of the work?

11-3. EVALUATING THE PROPOSAL

Sections 11-1 and 11-2 listed a number of benefits and costs for our proposed new control system. If the proposed system can show a net return that is significantly greater than the return for the existing system, then the proposed system would appear to be the better economic alternative. However, there may be still another alternative: Is this the best use of our money; can we and are we willing to invest that same money for a greater return in some way not connected with our present business? Also, is the plant investment worth the risk, which depends on many uncertainties? [46]

The soundness of our decision is vitally dependent on how accurate our estimates of benefits and costs turn out to be. When considering the economic climate of the future, a good deal of educated guesswork, crystal ball gazing, and finger crossing may be required. Yet, despite the uncertainties, a conscientious estimate is better than none at all and better than working by hunch for those of us who are not certified extra-sensory perceptionists. A poor decision may result in serious consequences, such as the following:

(1) *High Cost-to-Benefit Ratio.* The costs may be unnecessarily high compared to the benefits received. This may apply to the capital investment, the annual costs, or both.
(2) *Poor Performance.* The plant performance may be inferior, again resulting in financial consequences.
(3) *Bad Side Effects.* There may be adverse effects on the design or operation of equipment or the process in other parts of the plant.

(4) *Possible Lock-In.* Assume that a poor choice of system was purchased and must be paid for regardless of whether the system is kept in service or gotten rid of at once. Because the purchase cannot be undone, the investment is known as a *sunk cost.* All things considered, it may then be less costly to live with the poor choice from here on than to discard it and replace it. In such a case, the poor decision is irrevocable in practical terms.

Appendix A

Types of Instrumentation Documents

The systematic development of instrument systems for a specific plant project requires certain documents to be prepared. Most of these documents are prepared or initiated by instrument engineers, while some are developed by other disciplines and are used also by the instrument engineers.

The documents enable the instrument engineers to do the following:

(1) List their work as it progresses.
(2) Communicate with their co-workers, their client, manufacturers and their representatives, and construction people.
(3) Provide reference documents for themselves, the field people, and the plant engineers, operators, and service people.

Different organizations doing the same kind of work operate more or less differently from each other, and their paperwork may be somewhat different. However, in one way or another, they perform essentially the same functions.

For repetitive details of design, organizations normally simplify their work by making use of standards that they, themselves, produced, plus national standards published by ISA and other engineering societies. A *standard*

is a model or set of rules that define the requirements for a design, performance, or procedure for a specific purpose.

The use of standards speeds up the work because the answers to many questions concerning the subject matter of the standards are already there. At the same time, their use generally improves the quality of the work because each standard normally receives critical review before becoming official. In the case of national standards, such as those of ISA, the standards represent a consensus of good industry practice.

The engineering documents that instrument engineers produce are typically as follows.

PLANT PROCESS DIAGRAMS

Process Flow Diagrams

These are drawings to define the process. They are prepared by the process engineer, who is the person who designs the process in cooperation with the instrument engineer regarding matters of measurement or control. The process is the sole reason for having the plant; it is the beginning point of the project and the basis for requirements for land, energy, types and sizes of equipment, and controls. It is also valuable after the plant is operating.

The drawing shows the major process equipment and streams; and the fluid composition, pressure, temperature, flow, and other essential operating conditions at different places and for different cases of operation. The drawing shows the barest minimum of instruments to indicate which process variables need to be controlled or manipulated.

Piping-and-Instrument Diagrams (P & IDs)

The P & ID is an expanded version of the process flow diagram but without the detailed data on operating conditions. It shows all the piping, major and minor — including valves and other devices; all the equipment for the process; and utility systems for the plant — air, steam, fuel, and others. That much is handled by the process engineer.

The P & ID also includes all the instruments that the process requires, including minor instruments, such as thermometers and pressure gages. Incidental instruments, such as air-supply regulators for individual instruments, are generally not shown. The instruments are added to the drawing at the direction of the instrument engineer.

If a process subsystem — for example, a water-treating system — is purchased as a package, then the P & ID may represent it by a rectangle that shows no process equipment or instruments. Instead, the box carries a reference to a manufacturer's drawing that shows all the details. An alternative is to draw the package system details on the P & ID in order to complete the picture there.

The piping designer uses the P & ID to tell him schematically where to place instrument connections and straight runs of piping when needed for measurement.

The diagram is also used by many disciplines and by the operating company as an overall functional picture of the plant for use in discussions, for reference by the operators, maintenance people, etc.

INSTRUMENT CRITERIA

Instrument Application Criteria

The instrument engineer for a particular project prepares a document that states the guidelines that should generally be followed for selecting instruments. In effect,

the guidelines become project standards to promote uniformity and quality. They include items such as:

(1) Standard signal ranges.
(2) The types of sensors that should generally be used and limitations on their use.
(3) The variables that are to be put into the plant computer.
(4) The type of annunciator, size of windows and their color scheme, and the alarm sequence.
(5) The size and color of instrument cases to be mounted in the control room.
(6) Preferred brands of instruments. These may be based on the instrument engineer's experience and judgment but may also reflect the operating company's preferences.
(7) Many other points of instrument application philosophy and practice.

Instrument Installation Criteria

This document is prepared by the instrument engineer for a specific project to establish the guidelines to be generally followed for installing instruments. The document covers requirements such as the following:

(1) Root valves and instrument valves — their type and size.
(2) Sensing lines — their sizes, bleed connections, test connections, and general arrangement.
(3) Tubing materials, and the size and type of tube fittings.
(4) Installation tolerances.

LOOP DIAGRAMS

A loop diagram shows the detailed interconnections of the instruments of a loop and the instrument connections to power supplies. [55]

LOGIC CONTROL DIAGRAMS

These diagrams define the on-off functions of instrument systems, as covered in Chapter 6.

INSTRUMENT SPECIFICATIONS

These documents state the descriptions and requirements for instruments, instrument systems, and auxiliary equipment and services — such as startup assistance by the instrument manufacturer — that are to be purchased. ISA Standard S20 [83] provides 40 different fill-in specification forms for process measurement and control instruments. These forms are also available as computer software.

COMPUTER DOCUMENTS

A plant computer needs supporting documents that the instrument engineer prepares, such as the following.

(1) *Computer Input/Output List.* This list includes the following:

   (a) The input signals, and for each: its source, computer address, type —analog, binary, or digital — and its range and engineering units.

   (b) The functions that the signals are used for: for example, signal conversion, calculation, routine or abnormal display, or alarm.

   (c) The output signals, and for each: its computer address, destination, type — analog, binary, or digital — and its range and engineering units.

(2) *Computer Logic Diagrams.* A plant computer is customarily purchased with all its programming done by the supplier according to operating requirements outlined by the instrument engineer on computer logic diagrams. The programming defines the way that the computer carries out the control and information functions that would otherwise be carried out by conventional instruments.

(3) *Computer Display Diagrams.* The computer video displays may be of lists, statements, pictures, or a combination of these. Information for these displays is developed by the instrument engineer and given to the supplier for programming.

(4) *Computer Wiring Diagrams.* The instrument engineer prepares wiring diagrams or lists to enable the electrical designer to plan the wiring between the computer system, the related process instruments, and the power supply. The diagrams identify the computer system's electrical terminals, and show the devices to which they are wired.

ANNUNCIATOR DIAGRAMS

These show the window arrangement for the annunciator, a list of windows, their wiring points, and the sources of their input signals. If the annunciator has electrical outputs, these and their destinations are also listed.

CONSTRUCTION DIAGRAMS

Instrument Location Diagrams

These indicate the locations and elevations of instruments relative to the plant coordinates.

Instrument Board Diagrams

These show the outlines of the instrument boards, how they are sectioned, and how the instruments and other devices are laid out.

Instrument Installation Diagrams

These show the piping or tubing for connecting instruments to the process, to other instruments, and to pneumatic power supplies. They also describe methods of supporting instruments, and give other details of installation. They may also include instructions for special calibration to be performed in the field for individual instruments.

Instrument Heat-Tracing Diagrams

These list the instruments and sensing lines that require heat tracing or other type of weatherizing, and the desired control temperature for each application. They may describe how the tracing should be applied.

PURCHASE REQUISITIONS

The instrument engineer prepares a requisition, which is a written request for the purchase of instrumentation or services. The requisition contains a reference to an atttached specification for the items, and to other descriptive material, if needed. A requisition is usually the basis for invitations to bid that are sent to suppliers. After the instrument engineer reviews the bids, he may revise the specification and requisition. The purchasing department then issues a purchase order to the successful bidder.

SYSTEM DESCRIPTION

This is a description of the design and operation of the plant process, prepared by the process engineer, with additional material written by the instrument engineer and others. The instrument engineer's contribution is to discuss the instruments and how they are used to control the process.

The system description is a valuable reference document for all people who get involved in the working of the plant. It describes the plant as it was when built and should be kept up to date if changes are made.

INSTRUMENT INDEX

This is a tabulation that the instrument engineer prepares to keep track of all the instruments and their related documents. In its pure form, the index provides no direct information about the instruments but is merely a directory that points out where the information may be found. Substantively it says very little, yet it is a major work document for the instrument engineer and others during design, construction, and operation phases of the plant.

The initiation, expansion, and use of the index begin with the preparation of the piping and instrument diagram (P & ID) near the start of the project for work on a plant and continue throughout the entire project. Listing an instrument in the index before doing any engineering work on it helps to ensure that it is not later overlooked and that identification numbers are not duplicated.

For a large project, the index may consist of hundreds of pages of 15-inch-wide computer paper. The index is now usually computerized, thus making it easier to obtain lists of instruments that remain to be purchased, counting instruments of a given group, etc.

The tabulation in the index may include column headings such as those listed below. Each listed instrument has information supplied under each heading, as appropriate and as the information becomes available. The documents described above are included in the following list.

- Instrument identification number, for example, FE-1203. If instruments are provided and identified by a system-package manufacturer, then the manufacturer's instrument identification numbers may be listed.
- Instrument service, for example, "product to storage".
- P & ID number.
- Specification number.
- Purchase order number, if different from the specification number.
- Loop diagram number.
- Logic control diagram number.
- Computer input/output list number.
- Computer logic diagram number.
- Computer display diagram number.
- Computer wiring diagram number.
- Electrical schematic diagram number. This diagram is generally made by an electrical designer on the basis of the logic control diagram prepared by an instrument engineer.
- Annunciator drawing number.
- The number of the process piping or equipment drawing that shows the sensor connection. These drawings are generally made by a piping designer or an equipment designer, respectively.
- Instrument location diagram number.
- Instrument installation drawing number.
- Heat-tracing list number.
- Instrument board drawing number.
- The numbers of the manufacturer's documents, such as outline drawing, wiring diagram, installation manual, service manual.
- Remarks.

Appendix B

Gas versus Vapor

The words *gas* and *vapor* are not synonymous, although they are sometimes used loosely and interchangeably. The difference between them is described below.

When a pot of water is boiled, the liquid evaporates and becomes steam. The steam is an invisible vapor. Steam can usually be condensed back to the liquid state by either cooling or compressing it. If we see a mist or fog coming from the water in the pot, we still do not see the steam. What we see is fine droplets of water that condensed when the steam met cool air.

If the water vapor is not cooled but is compressed by a piston inside a cylinder, steam condenses because of the increased pressure. If we put the vapor in a heater that raises the temperature above 705.40°F (374.11°C), the steam cannot condense no matter how much it is compressed. This temperature is known as the *critical temperature* of water. Above the critical temperature, water is a *gas*; below that temperature, it is a *vapor* or, if it gets cold enough, it is a *liquid* or solid *ice*.

Every substance has its own critical temperature above which it is impossible to liquefy the substance and above which the substance is a gas. Examples are:

Substance	Critical Temperature	
Nitrogen	−233°F	(−147°C)
Oxygen	−182°F	(−119°C)
Methane	−117°F	(−83°C)
Propane	206°F	(97°C)
Ammonia	270°F	(132°C)
Chlorine	291°F	(144°C)
Ethyl Alcohol	470°F	(243°C)

Pure air that we breathe consists almost entirely of nitrogen and oxygen, which are gases so far as we are concerned because they are normally far above their critical temperatures. Propane, the so-called *LP gas*, sold in bottles for use in portable stoves, actually exists as a vapor at room temperature. Gasoline fumes are vapors. Alcoholic drinks release ethyl alcohol vapor into the air.

121

References

1. Instrument Society of America (ANSI/ISA) Standard S5.1-1984, *Instrumentation Symbols and Identification*, Section 2, *process*
2. ANSI/ISA Standard S5.1, Section 2, *process variable*
3. ANSI/ISA Standard S5.1, Section 2, *measurement*
4. ANSI/ISA Standard S5.1, Section 2, *controller*
5. ANSI/ISA Standard S5.1, Section 2, *final control element*
6. ANSI/ISA Standard S5.1, Section 2, *loop*
7. ANSI/ISA Standard S51.1-1979, *Process Instrumentation Terminology*, Section 4, *range*
8. ANSI/ISA Standard S51.1, Section 4, *span*
9. ANSI/ISA Standard S51.1, Section 4, *time constant*
10. ANSI/ISA Standard S51.1, Section 4, *calibrate*
11. ANSI/ISA Standard S51.1, Section 4, *error*
12. ANSI/ISA Standard S51.1, Section 4, *accuracy*
13. ANSI/ISA Standard S51.1, Section 4, *modulation*
14. ANSI/ISA Standard S51.1, Section 4, *signal, digital*
15. ISA Standard S7.4-1981, *Air Pressures for Pneumatic Controllers, Transmitters, and Transmission Systems*
16. ISA Standard S50.1-1982, *Compatibility of Analog Signals for Electronic Industrial Process Instruments*
17. American National Standards Institute/Institute of Electrical and Electronics Engineers (ANSI/IEEE) Standard 268-1982, *Metric Practice*, Appendix *A*
18. ISA Recommended Practice RP3.2-1978, *Flange-Mounted Sharp-Edged Orifice Plates for Flow Measurement* (withdrawn)
19. *Fluid Meters: Their Theory and Application*, Sixth Edition, American Society of Mechanical Engineers, 1971
20. ANSI Standard MC96.1-1982, *Temperature Measurement Thermocouples*, sponsored by ISA
21. ANSI/ISA Standard S51.1, Section 4, *set point*
22. ANSI/ISA Standard S51.1, Section 4, *gain, proportional* and *proportional band*
23. G. K. Tucker and D. M. Wills, *A Simplified Technique of Control System Engineering*, Appendix *B*, Minneapolis-Honeywell Regulator Company, 1962
24. ANSI/ISA Standard S51.1, Section 4, *noise*
25. ANSI/ISA Standard S51.1, Section 4, *control, cascade*
26. ANSI/ISA Standard S51.1, Section 4, *signal selector*
27. ANSI/ISA Standard S51.1, Section 4, *signal amplitude sequencing (split ranging)*
28. ANSI/ISA Standard S51.1, Section 4, *control, feedforward*
29. ANSI/ISA Standard S51.1, Section 4, *control valve*
30. ANSI/ISA Standard S5.1, Section 2, *control valve*
31. ANSI/ISA Standard S51.1, Section 4, *hunting*
32. R. W. Miller, *Flow Measurement Engineering Handbook*, First Edition, 1983, McGraw-Hill Book Company
33. ANSI/ISA Standard S75.01-1985, *Flow Equations for Sizing Control Valves*
34. J. W. Hutchison, *ISA Handbook of Control Valves*, Second Edition, 1973, Instrument Society of America
35. ANSI/ISA Standard S5.1, Section 6.7

36. National Electrical Manufacturers Association (NEMA) Standard ICS 3-1978, Part ICS 3-304, *Industrial Systems*

37. Scientific Apparatus Makers Association (SAMA) Standard PMC 22.1-1981, *Functional Diagramming of Instrument and Control Systems*

38. U.S.Bureau of Mines Bulletin 503

39. National Fire Protection Association (NFPA) Publication 70-1981 (ANSI C1-1981), *National Electrical Code 1981*; and the equivalent Canadian Standards Association Standard (CSA) C22.1-1982, *Canadian Electrical Code Part I*

40. ISA Recommended Practice RP12.1-1960, *Electrical Instruments in Hazardous Atmospheres*

41. ISA Standard S12.10-1988, *Area Classification in Hazardous (Classified) Dust Locations*

42. ISA Standard S12.11-1973, *Electrical Instruments in Hazardous Dust Locations*

43. American Petroleum Institute Publication (API) 500A, 3rd Edition, April 1966, (reaffirmed 1973), *Recommended Practice for Classification of Areas for Electrical Installations in Petroleum Refineries*

44. ISA Standard S12.4-1970, *Instrument Purging for Reduction of Hazardous Area Classification*

45. ISA Recommended Practice RP12.6-1987, *Installation of Intrinsically Safe Systems for Hazardous (Classified) Locations*

46. Grant, E. L., Ireson, W. G., and Leavenworth, R. S., *Principles of Engineering Economy*, Seventh Edition, John Wiley and Sons, New York, 1982

47. ANSI/ISA Standard S51.1, Section 4, Reliability

48. IEEE Standard 279, *Criteria for Protection Systems for Nuclear Power Generating Stations*

49. IEEE Standard 379, *Guide to the Application of Single-Failure Criterion to Nuclear Power Generating Station Protection Systems*

50. ISA Recommended Practice RP60.3-1985, *Human Engineering for Control Centers*

51. United States Department of Defense (DOD) Standard MIL-STD 14728 and Addenda, *Human Engineering Design Criteria for Military Systems, Equipment, and Facilities*

52. ISA Standard S5.1, Section 2, *pilot light*

53. ISA Standard S5.1, Section 2, *alarm*

54. ISA Standard S18.1-1979 (R 1985), *Annunciator Sequences and Specifications*

55. ANSI/ISA Standard S5.4-1976 (R 1981), *Instrument Loop Diagrams*

56. ANSI/ISA Standard S5.2-1976 (R 1981), *Binary Logic Diagrams for Process Operations*

57. ANSI/ISA Standard S5.1, Section 2, *transmitter*

58. L.K. Spink, *Principles and Practice of Flow Meter Engineering*, Ninth Edition, 1967, The Foxboro Company

59. American Gas Association (AGA), *Gas Measurement Manual*

60. ANSI/ISA Standard S5.1, Section 4.2

61. ANSI/ISA Standard S5.1, Section 5, Table 3, *high-selecting* and *low-selecting*

62. Norman A. Anderson, *Instrumentation for Process Measurement and Control*, Third Edition, 1980, Chilton Company

63. Gregory K. McMillan, *Tuning and Control Loop Performance*, 1983, Instrument Society of America

64. American Society of Mechanical Engineers (ASME) Boiler and Pressure Vessel Code

65. ANSI Standard Y32.2-1975/Canadian Standards Association (CSA) Standard Z99-1975/IEEE Standard 315-1975, *Graphic Symbols for Electrical and Electronics Diagrams*

66. ISA Standard S71.01-1985, *Environmental Conditions for Process Measurement and Control Systems: Temperature and Humidity*

67. ISA Standard S71.04-1985, *Environmental Conditions for Process Measurement and Control Systems: Airborne Contaminants*

68. NEMA Publication No. 1S 1.1-1975

69. ANSI/ISA Standard S7.3-1975 (R 1981), *Quality Standard for Instrument Air*

70. American Nuclear Society (ANS) Standard 59.3-1977, *Safety-Related Control Air Systems*

71. Carl L. Moore and Robert K. Jaedicke, *Managerial Accounting*, Chapter 19, 2nd Edition, South-Western Publishing Company, 1967

72. American Gas Association (AGA), *AGA Gas Measurement Manual*

73. ISA Standard S12.12-1984, *Electrical Equipment for Use in Class I, Division 2 Hazardous (Classified) Locations*

74. ANSI/ISA Standard S5.5-1985, *Graphic Symbols for Process Displays*

75. ANSI/ISA Standard S51.1, Section 4, *repeatability*

76. ANSI/ISA Standard S51.1, Section 4, *reproducibility*

77. Theodore Baumeister, *Marks' Standard Handbook for Mechanical Engineers*, Seventh Edition, 1967, McGraw-Hill Book Company

78. NEMA Publication No. 1S 1.1-1975

79. ISA Standard S5.3-1983, *Graphic Symbols for Distributed Control/Shared Display Instrumentation, Logic and Computer Systems*

80. ANSI/ISA Standard S51.1, Section 4, *control mode*

81. ANSI/ISA Standard S51.1, Section 4, *control action*

82. ANSI/ISA Standard S51.1, Section 4, *direct-acting controller* and *reverse-acting controller*

83. ISA Standard S20-1981, *Specification Forms for Process Measurement and Control Instruments, Primary Elements and Control Valves*

Index